Goat Health and Welfare

A VETERINARY GUIDE

Goat Health and Welfare

A VETERINARY GUIDE

David Harwood

THE CROWOOD PRESS

First published in 2006 by
The Crowood Press Ltd
Ramsbury, Marlborough
Wiltshire SN8 2HR

www.crowood.com

British Library Cataloguing-in-Publication Data
A catalogue record for this book is available from the British Library.

ISBN 1 86126 824 6
EAN 978 1 86126 824 2

Disclaimer
The author and the publisher do not accept responsibility in any manner
whatsoever for any error or omission, nor any loss, damage, injury
or liability of any kind incurred as a result of the use of any of the
information contained in this book, or reliance on it.

Acknowledgements
To the many goat keepers who have allowed me to investigate problems
in their goats, our conversations have taught me a lot. To the many
veterinary practitioners who have shared their goat problems with me.
To members of the British Goat Society for nominating me as one of
their honorary vets and thus stimulating my interest, and to members of
the Goat Veterinary Society for helpful discussions.

Most of the photographs are my own, although I would like to
acknowledge those provided by: Martin Andrews, Kathy Anzuino,
Cornelia Bidewell, the British Goat Society, Tom Clarke, Stephen Cobb,
Kate Dedman, John Fishwick, Jane Marshall, Eamon Watson, VLA
Winchester and Agnes Winter.

Apologies to anyone whose photograph I have inadvertently used and
not acknowledged – please accept my belated thanks.

Dedication
To my grandfather Albert Allen for his original inspiration.

To my mother Joyce and my late father Malcolm for their constant
support throughout my professional and personal life.

Last but not least to my wife Gerry, and to Maddy, Jessica, Nicola and
Kimberley for all their support and encouragement and for continuing to
put up with me!

Typeset in Century Schoolbook by Bookcraft Ltd, Stroud,
Gloucestershire

Printed and bound in Great Britain by CPI Bath

Contents

Foreword

Goat keeping in the UK is undergoing great change as we adapt to new EU legislation and the ever-increasing Rules and Regulations of our own Government Agencies. The 2001 Foot and Mouth outbreak ensured that future health and welfare issues would be high up on the agenda for all livestock breeders.

Commercial goat keeping is increasing in the UK as consumers recognize the quality of goat produce. Large units have developed in order to meet the demand for dairy goat products. Meat and fibre goats have also enjoyed an increase in popularity and Pygmy goats are becoming far more common as people recognize their attraction as pets. Goats are still kept in small numbers to produce milk for the household, and exhibition stock continue to demonstrate the successful breeding and improving of goats in the UK during the last 125 years.

The goat is often described as the 'Universal Foster Mum' as her milk is used successfully to rear the young of many species. The small fat globules in goats' milk make it particularly easy to digest. Goats' milk can also make a suitable alternative for those who have an allergy to bovine products.

Good husbandry and stockmanship result in an understanding of our animals and recognition of the signs that tell us when all is not well. It is only by knowing how a healthy animal behaves that we recognize when problems arise and things go wrong.

David Harwood's book is written in layman's language and is easy to read and easy to understand. It is thorough and well illustrated throughout, and should become an essential addition to any goat keeper's library. Chapters on goat behaviour, feeding and management, recognizing signs of ill health and on prevention being better than cure make this an invaluable practical reference for goat keepers and anyone involved in goat husbandry.

David Harwood is one of the British Goat Society's Honorary Vets and spends a great deal of time visiting Veterinary Colleges throughout the UK lecturing students on goat health and welfare issues. He is also a much sought-after speaker at British Goat Society Conferences. The work that David Harwood undertakes plays an important role in promoting the aims of the British Goat Society and in goat keeping in general.

I am delighted to be invited to write this foreword to *Goat Health and Welfare – a Veterinary Guide*, and I recommend it to anyone who has the best interests of this useful and endearing animal at heart.

Richard Wood
Chairman – British Goat Society

Preface

Goats are kept for a wide variety of reasons around the world, and this makes the species unique in the animal kingdom. They are ruminants like cattle and sheep, and as a result are farmed for milk, meat and fibre. On many units they are kept intensively, with large herds housed all year round on specialist rations. Conversely, however, many goats are kept in very small groups purely as a hobby, providing a small quantity of milk for liquid consumption by the household, or to be made into cheese or yoghurt. Yet a further group of owners keeps single goats, or small groups of, for example, pygmy goats purely as pets.

The goat is able to adapt to this wide variation, but relies heavily on its owner or keeper for its well-being, no matter how it is kept.

I have based this book mainly on my own experiences, and have tried to give guidance to the reader on how to ensure that goats in their care remain fit and healthy, emphasizing the importance of a thorough knowledge of, familiarity with, and respect for the species. But inevitably things will go wrong, and a goat may become ill or injured. Such problems need to be recognized promptly, and guidance is provided throughout the book on how to achieve this. It is important, however, that the reader recognizes his, or her, limitations. This is particularly the case when professional help needs to be sought – a good working relationship with your vet is an important one to develop.

If you are establishing a new goat unit, or even contemplating buying a goat as a pet, make adequate time to prepare both your unit and yourself! Look out for local sources of information, join a local goat club, visit local goat units, and ask your vet to come and visit you when the goats have arrived, establishing an early working partnership.

Goats are friendly, inquisitive and fascinating creatures, and they will give endless pleasure to their keepers. Notwithstanding this, I have become concerned in the past few years about the number of inexperienced owners keeping goats as pets, and some of the resulting problems that I have encountered.

I hope this book will go some way to ensuring that the health and welfare of all goats is not compromised, and that those keeping them – for whatever reason – continue to find them rewarding and stimulating, often frustrating, but always an interesting species.

David Harwood
September 2005

Introduction

This book has been written to give the novice goat keeper some help in ensuring that the goat or goats that they keep, remain fit and healthy. It is lavishly illustrated throughout, and therefore should also be a useful reference guide to more experienced goat keepers, including those who farm goats commercially on a large scale.

The book begins by considering normal goat behaviour, in particular with a view to recognizing early signs of ill health often manifested by subtle behavioural changes. The value of becoming a good stockperson is the key to success, and this is a constant theme. Normal goat activity such as feeding and nutrition is considered, and how goats adapt to management change associated with housing, grazing and so on. Signs of ill health are discussed in general terms to give the reader some further pointers to recognize signs of disease, and the remaining chapters then consider each of the major organ systems, and the health and welfare problems that can arise.

Throughout the text the importance of a good working relationship with the reader's own vet is emphasized, with pointers given as to when a vet needs to be called or consulted.

Goats are fairly adaptable creatures in the wild, but if they are kept as pets or farmed, they rely very heavily on their owners for food, and for protection from disease, injury or predators. And as the book title suggests, goat health and welfare are closely linked in that a fit, healthy goat is a 'happy' goat; its welfare is not compromised, and if it is farmed commercially, it will also remain productive.

CHAPTER ONE

Goat Behaviour

Goats, when content and in good health, are usually quiet creatures of habit, full of energy, often playful and impulsive. A good stock-keeper should be able to recognize when a goat is ill, often simply by observing its behaviour and by noticing changes in its normal behavioural activities. If well looked after, there is no reason why a goat should not reach twenty years of age.

It is not often that you will hear a goat bleating or vocalizing for no reason, and if they begin to make a noise it is often a sign that something may be wrong. There are exceptions to any rule, however, and some goats make gentle noises when they are content, and become quieter when something is wrong. Knowing your stock is vitally important.

Good stockmanship is the key to keeping all goats fit and healthy, and the earlier a problem is recognized, the more likely it is that any treatment administered will be effective. This is particularly true when goats are kept in large numbers, since these early subtle changes may be more difficult to recognize. In large groups, however, there are other signs that indicate that a goat may not be feeling well, such as a reluctance to enter the milking parlour for milking, hanging back from feeding troughs, and standing in the background away from the main group activities.

There are three reasons why a normally quiet, placid goat may start bleating, and in particular bleating more constantly:

- When it is hungry or thirsty, and this is usually alleviated by giving food/water, resulting in a quick return to normal behaviour.
- During the breeding season, particularly when a doe is on heat, or during the latter part of pregnancy as kidding approaches. There may be a similar change in activity during weaning.
- When a goat is sick or in pain it will often begin to vocalize, and as with the human voice, the volume and pitch will change depending on the intensity of pain or discomfort felt. Conversely, as the animal deteriorates, or in a goat that is feeling really miserable, the sound will be of a lower intensity.

Goats seem to have a relatively low pain threshold when compared with some other farm animals, such as cattle, and they don't tolerate ill health very well. A diligent owner will be aware of this, as also the need to give plenty of 'TLC' (tender loving care) to any sick goat to ensure that it continues to feed, is warm and comfortable, and that it maintains an interest in life.

A HEALTHY GOAT

In addition to the behavioural features already discussed, goat owners should familiarize themselves with the physical appearance of a healthy goat, then with experience they will recognize any subtle changes that may indicate that all is not well.

The head should be bright and alert.

The head: The head is normally held up with a bright, alert attitude. It is not a good sign if it is lowered and the goat is showing little interest in what is going on around it.

The eyes: These should be bright and kept open; look out for any discharge, or for a constant blinking, as this may indicate a problem. By gently lifting one eyelid, look at the colour of the conjunctiva: in the healthy goat it is pink, if it is pale the goat may be anaemic.

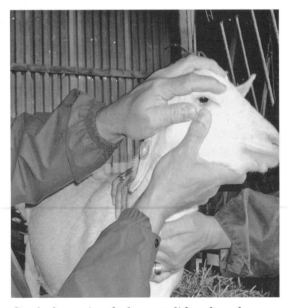

Gently depressing the lower eyelid to show the mucous membranes.

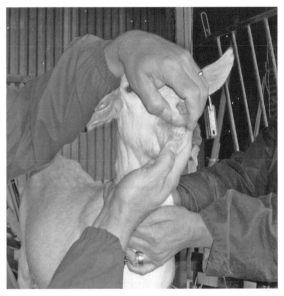

A general examination of the mouth and teeth.

The ears: Ear shape, size and carriage will vary among goat breeds; thus some will hold their ears erect, whereas others – for instance Anglo Nubian goats – will maintain a droopy ear. Again, a good stockman will recognize any change from the normal appearance. A swollen ear (haematoma or blood blister) may occur if the head is being constantly shaken due, for example, to ear irritation caused by mange mites.

Mouth and teeth: Goats are usually very clean feeders, and are selective and careful what they eat. It follows, therefore, that the mouth should be clean, and there should be no saliva drooling from the lips. There should be no swellings along the jaws or cheeks, or any discomfort when the mouth is handled. Watch your goats 'chewing the cud', and get used to the actions involved. Many goats will occasionally grind their teeth for no real reason, but excessive teeth grinding can indicate pain.

Respiration: Again, it is important to observe your goats, and to get used to their normal breathing rate. This will vary tremendously depending, for example, on whether they are resting or have just run in from the field. An apprehensive or frightened goat will often breathe more quickly. The respiratory rate will also be higher on hot days than on cold. It follows therefore, that although the rate will rise in illness, particularly in a goat with pneumonia or a high temperature, the worried owner should always look for other signs of illness. In extreme respiratory distress, the breathing rate will be faster, but the goat will also hold its head out straight to allow air to pass freely up and down the airways, and its nostrils will be flared. These are signs that all is not well; look out for an accompanying cough.

The respiratory rate in adults is 15 to 30 per minute, and in kids it is 20 to 40 per minute.

Temperature: The normal goat temperature is again quite variable, depending on whether it has been recently stressed, for instance by gathering or if it is a hot day. It is generally accepted that the temperature should be 38.6–40.6°C (102°–104°F), average 39.3°C (103°F). A thermometer can be a useful part of your first-aid kit or medicines cabinet; you can use either a stub-ended glass mercury thermometer or a thermometer with a digital read-out. The goat's temperature is taken 'per rectum', when the end of the thermometer is gently inserted into the anus, using a small amount of lubricant for ease of insertion. Remember to shake down the mercury in a glass thermometer before use, or to re-set the dial in a digital thermometer. For baby goats the easiest way is to lay the kid across your lap; taking the temperature of a larger goat is usually a two-person job, and is best done with the animal standing. Leave the end of the thermometer inside the rectum for at least two minutes before you remove it, then wipe it clean and record the temperature. If you are concerned about your goat running a high temperature, then phone your vet and discuss it, remembering to record any other observations you may have made.

Pulse/heart rate: The pulse rate of an adult animal varies depending on its size, but normally it falls within the range 65–95 per minute. As with temperature and respiratory rate, this will rise if the goat has been stressed

The thermometer is gently inserted into the anus, after lubrication.

or moved, or on hot days. It isn't easy to take a goat's pulse, and normally the heart rate gives the best indication. A veterinary surgeon will usually listen to the latter using a stethoscope, but you can get a fairly good indication by holding your hand over the heart area (placing your hand over the chest wall just inside the elbow).

Locomotion: Goats will normally graze or feed on all four feet, with the feet placed squarely on the ground. Although some goats will occasionally feed with their knees bent (and you should recognize those that do), if a goat suddenly adopts this stance it may be an indication that it is lame, usually in one of its front limbs. You may also notice hair loss and thickening of the skin on the front of the knee joint. Watch your goats moving: a sound animal will normally have a well defined and regular gait, picking a spot to place its front foot, with the back foot following in more or less the same spot as it moves forwards. If it is lame, then the gait becomes irregular, often with a hopping movement, and when stationary the painful limb will often be held off the ground, or with the tip of the toe just touching the ground.

Skin and hair covering: Although hair covering will vary depending on the type and breed of goat (from the fine hair covering of the traditional dairy breeds to the thick

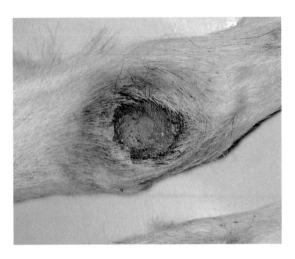

This goat has been feeding on its knees (carpal joints); note the hair loss that has occurred.

covering of the Angora), there are a number of signs of ill health to watch out for. All goats will moult, particularly when warmer weather arrives, but an excessive moult may be cause for concern, particularly if only a single goat is affected. A goat in poor bodily condition (for whatever purpose) may moult excessively, and the hair may become fragile and break easily. Look for other signs – is the hair loss generally all over the body, or is it confined to just one or more areas? Is the hair loss symmetrical? This may indicate a hormonal problem. Is there any sign of itchiness, or scab and scurf formation? A goat exhibiting excessive rubbing, scratching or nibbling is usually fairly distressed, and should be dealt with urgently, by seeking veterinary attention to determine a cause.

Digestive tract: The ruminant digestive tract is a complex structure, and its function will be discussed in more detail in Chapter 7. There are, however, a number of features that an experienced stockman will recognize that indicate a good functional digestive system, and conversely others that he will recognize when things are going wrong. An integral part of the ruminant digestion is the regurgitation of the contents of the rumen back up into the mouth for further chewing. This process is known as rumination, or 'chewing the cud', and normally takes place when the goat is

relaxed and recumbent. Watch the procedure: you can follow the cud as it moves up the neck, and the goat will then commence a chewing routine with jaw movements to left and right, and at a remarkably consistent rate and duration, before swallowing.

Cessation of rumination is usually a sign of ill health, and should be recognized. Belching regularly expels gas that builds up in the rumen, and this again is a normal procedure. If this mechanism fails – and there are a number of reasons why it may – then gas will build up in the rumen, causing the abdomen to become distended and visible, initially high on the left side, where the normally 'hollow' flank will fill out. If pressure begins to build, the abdomen will become distended and progressively tighter, and the goat will become very agitated; in this condition it is said to be bloated, and this is a true veterinary emergency when help should be sought quickly.

Become familiar with the appearance of the droppings. Although they should normally be pelleted, it is not unusual for the faeces to become looser and more 'pasty', particularly when the goat is at grass. However, it is the sudden change to loose or watery faeces – so-called 'scouring' – that may indicate a developing gut problem. If it is excessive or blood-tinged, then consult your veterinary surgeon without delay, as it may indicate a serious problem. Diarrhoea in young kids can be very debilitating, and can quickly cause dehydration and even kidney failure if untreated. Excessive straining to pass faeces may also be cause for concern; it could indicate a blockage, or a severe parasite infection such as coccidiosis.

Reproductive tract: Goats are seasonal breeders, and the onset of cyclical activity in does (in the northern hemisphere) is usually between the months of August and February, with the greatest activity in the period September to December. The stimulus to breed is a declining day-length, and many commercial units now use this phenomenon to encourage out-of-season breeding using artificial lighting.

The normal doe is on heat (oestrus) for

A normal buck – don't forget him!

approximately thirty to forty hours, although this may extend up to ninety hours. Oestrus may result in much vocalization, and this behaviour can be confused with a goat in pain or distress; however, it is quite normal.

Never forget the buck – he needs to be fit and healthy if he is to be fertile during what can be a relatively short breeding season.

Urinary system: Watch your goats urinate, and familiarize yourself with the attitudes and postures adopted. A normal healthy female will crouch slightly to pass urine, and may adopt this attitude and pass urine when she is apprehensive – it is a normal behavioural activity. The male will often pass urine in dribbles rather than in a steady stream, and entire bucks may well urinate over their limbs and head, to enhance their hormonal attraction! Neither males nor females should have any difficulty in passing urine, and straining, particularly in males, is an indication that something is wrong. Urine colour will vary depending on a number of factors, but should be a pale yellow 'straw' colour. Very dark urine may indicate dehydration and illness.

Goats should have access to clean water at all times; they are relatively fastidious, and may refuse to drink if the water is dirty or contaminated. Failure to drink in males (particularly castrated males) may encourage stone formation in the bladder.

Body condition: Up to 25 per cent of a goat's bodyweight can be attributed to the weight of food in its rumen; it is also recognized that a 5–10 per cent fluctuation in bodyweight can take place in a three-day period via variation in feed intake alone! Although weighing is a useful means of assessing body condition (particularly when administering medication), readers should familiarize themselves with the concept of 'body condition scoring' (*see* Chapter 2).

Animal behaviour: Interest in the science of animal behaviour, and particularly its association with the health, welfare and productivity of farm livestock, has been increasing in recent years. There are a number of other more general points worth emphasizing when keeping goats:

- Goats are gregarious animals, can become easily bored, and will constantly look for

Environmental enrichment – this old chair makes an ideal plaything!

stimuli – hence their apparent desire to investigate anything in their environment, including light switches, electric cables, and so on! It is therefore worth considering some form of environmental enrichment, such as chains hanging from the ceiling, mounds of earth or artificial platforms such as straw bales to climb and clamber on. This is particularly important for goats kept singly as pets.

- There may be subtle breed differences. For instance, British Alpine, British Toggenburg and Saanen goats appear to be more 'laid back' and are not easily unnerved, whereas Anglo Nubian and Golden Guernsey goats are more easily unsettled, although as ever there will be individual exceptions.
- They are very hierarchical and will quickly develop their own 'pecking order'; once established this will be respected, but it can cause problems if new group members are introduced. It follows, therefore, that goat groups need to remain fairly static where possible, as constant movements and re-grouping can be unnerving, and may lead to reduced milk yields, and even predispose to illness. In a mixed sex group, it is rarely the male that is the dominant animal: invariably there will be a dominant female. Goats can also be 'politically insensitive' and move towards mixing with their own colour! For example, when bringing different groups of kids together, you may find that they will automatically team up with kids of the same type and colour.
- It is widely recognized that goats develop long-term 'friendships', particularly between litter sisters, coupled with a fairly long-term memory for each other even when separated for significant periods of time. A good stockman will recognize this, and will make sure that such friendships are not broken.
- Goats also appear to develop an affinity for owners, and are often unnerved by changes in owner and environment.

A behavioural phrase worth remembering for anyone keeping or handling goats is this: 'Get your goat to do what you want it to do, but in the way it wants to do it.'

CHAPTER TWO

The Normal Goat – Feeding and Management

BASIC NUTRITION

It is important to ensure that all goats receive a balanced diet that provides an adequate daily input of nutrients to fulfil basic maintenance requirements (that is, to maintain bodyweight and function). There may be additional requirements for pregnancy and lactation in adult does during the breeding season, for breeding bucks prior to and during the breeding season, and for body growth in young stock. Any ration fed should recognize this.

Food should be of good nutritional quality, and should be palatable and in a form the goat will eat. If feeding a rationed diet, then care should be taken to ensure that all goats can eat together – there should be sufficient trough or feeder space – otherwise dominant goats will eat more, leaving less for subordinate (shy) goats.

Like cattle and sheep, goats are ruminants, and it is important for the goat keeper to have an understanding of the ruminant digestive system so as to ensure that a suitable ration is fed, and thus avoid digestive upsets such as diarrhoea or bloat.

RUMINANT DIGESTION

The ruminant gut has evolved to enable ruminants to consume large quantities of plant material relatively quickly. It is thought that the process of 'rumination' (or cudding), whereby food is regurgitated and chewed later (usually during a quiet time of the day or night) and then swallowed again, evolved to enable them to eat large quantities of food quickly, then chew it properly when they are in a safer environment and not out in the open. It is suggested that in a group of goats at least two-thirds of their number should be cudding at any one time, and if this drops to a half or less, there might be a problem with, for example, acidosis (*see* Chapter 7).

The goat has four stomach compartments, the largest one being the rumen or first stomach. It is in the rumen that a large population of micro-organisms aid the breakdown and digestion of plant cell wall material

All goats should be able to feed together.

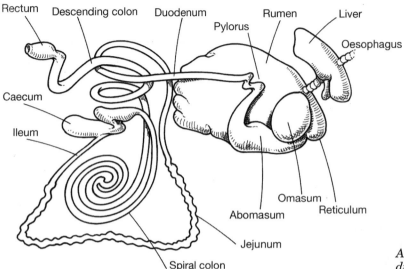

Rectum | Descending colon | Duodenum | Rumen | Liver
Pylorus | Oesophagus
Caecum
Ileum
Omasum | Reticulum
Abomasum
Jejunum
Spiral colon

A diagram of the ruminant digestive system.

(mainly cellulose), thus releasing the available nutrients: it has a constant movement, effectively mixing the food and saliva together. Any disruption of this microbial population of the rumen can have serious consequences. Any sharp fragments in the diet, such as stones, pieces of wire or nails and screws, will drop into the reticulum (the second stomach), thus protecting the more delicate fourth stomach (or abomasum) from damage. The third stomach, or omasum, effectively dries the rumen content by absorbing moisture via its leaf-like structure. The abomasum is the equivalent of the stomach of a single-stomach animal (for example, man), and allows food material already broken down in the fore-stomachs to be acted on by gastric enzymes. Further passage of the food through the gut stimulates the secretion of pancreatic juices and bile, both aiding digestion; and following absorptive processes in the gut, faeces will appear (normally pelleted).

Any abnormality in the procedures outlined above will result in an inadequate uptake of nutrients from the ration (so-called malabsorption) and loss of weight, and may be accompanied by changes in faecal consistency.

During rumen digestion, volatile fatty acids (VFAs) and methane gas are produced. VFAs are absorbed through the rumen wall and are the main source of energy; these are predominantly acetic and propionic acid with a smaller quantity of butyric acid (these will vary depending on the diet fed). An excessive intake of readily digestible material such as barley can result in excessive production of lactic acid, and the resulting acidosis can be fatal (*see* Chapter 7).

Any methane is removed by fairly regular eructation (belching) of gas, but an excess accumulation can be fatal, causing bloat to develop followed rapidly by death.

Goats will chew the cud when relaxed and resting.

NUTRITIONAL PRINCIPLES

It is beyond the remit of this book to discuss feeding in great detail, and the reader requiring further information is advised to consult other sources (including *Practical Goat Keeping* by Alan Mowlem, published by the Crowood Press).

Dry matter intake (DMI): Goats can only eat a certain amount of food each day, referred to as their 'dry matter intake' or 'DMI'. The water content or 'fresh weight' of a dietary component (for example, grass) can vary greatly, and by referring to its DMI when working out a suitable ration this can be over-ridden. As an example, 2kg (4.5lb) of fresh grass would be different to 2kg of hay, and different again to 2kg of barley. There are tables available giving the dry matter content of different feeds, and an example is given below (*see* Glossary for an explanation of the terms used).

The dry matter intake itself will vary from goat to goat depending mainly on its weight, but it is generally considered to be approximately 3.5 per cent of its bodyweight per day: that is, a 50kg (110lb) goat is capable of consuming approximately 1.75kg (3.75lb) DMI per day. This will rise, particularly in high-yielding milking goats, but it can be reduced in the latter stages of pregnancy, particularly when carrying multiple kids, thus reducing the available rumen volume within the abdominal cavity.

Energy: In most feeding guides the term 'metabolizable energy' (ME) is used to assess energy intake, usually in megajoules (MJ). The energy demands of goats will vary depending on their physical and physiological demands and level of production; however, as a rough guide:

- Maintenance: $0.5\text{MJME/kg}^{0.75}$ of metabolic bodyweight (liveweight in kg to the power of 0.75).
- Pregnancy: $0.5\text{MJME/kg}^{0.75}$ rising to $0.7\text{MJME/kg}^{0.75}$ for the last month.
- Lactation: maintenance needs + 5 MJME/ kg milk produced (assuming 3.5 per cent butterfat).

	Dry matter (%)	ME (MJ/kg DM)	Crude protein (g/kg DM)	RDP (g/kg DM)	UDP (g/kg DM)	Crude fibre (g/kg DM)
Silage						
Grass (good quality)	27	10.2	170	136	34	300
Maize	21	10.8	110	66	44	233
Hay						
Grass	85	9.0	101	81	20	320
Straw						
Barley	86	7.3	38	30	8	394
Wheat	86	5.7	24	19	5	426
Green crops						
Grass	20	11.2	175	105	70	225
Grains						
Barley	86	13.7	108	86	22	53
Oats	86	11.5	109	87	22	121
By-products						
Dried brewers' grain	90	10.3	204	122	82	169
Sugar-beet pulp	90	12.2	106	64	42	144
Maize gluten	90	14.2	394	315	79	23

Some typical nutritional values of a range of feedstuffs used for feeding goats

Physical factors also affect energy demands – for instance, environment (demands are higher in cold weather), or exercise (demands are higher at grass than housed).

Protein: Our understanding and application of protein nutrition has undergone many changes in the past few years, and this must be borne in mind when consulting older texts on this subject. The current system used for formulating protein requirements in the rations of ruminants applies the concept of 'totally absorbed amino acid nitrogen (AA-N)'; older texts may still refer to digestible crude protein (DCP), rumen degradable protein (RDP), and undegradable dietary protein (UDP). If in doubt, seek specialist nutritional advice; your veterinary surgeon may be able to help.

Minerals and trace elements: Although some specific health problems are referred to elsewhere in this book, it is fair to say that specific mineral and trace element deficiencies are not as common as in either cattle or sheep. Hypocalcaemia (milk fever) and hypomagnesaemia (grass staggers) are both comparatively rare compared to their occurrence in dairy cows, even in high-yielding Saanen milking goats. Copper deficiency can be a specific problem, particularly in its association with 'swayback' in newborn kids.

It seems likely that the selective feeding behaviour of the goat allows it to choose feed materials it may need. Given the opportunity, goats at grass will selectively graze weeds, branches or scrub and suchlike, which are much more mature than, say, fresh herbage. It follows, therefore, that if this natural browsing behaviour is over-ridden by, for example, housing, then a good, well balanced diet must be provided by the goat keeper, and many owners will bring in hedgerow cuttings or branches – assuming that these are suitable, and not poisonous!

Care should always be taken that a deficiency is not converted to toxicity by over-supplementation – for example, copper.

Water: A goat may drink up to 18ltr of water (4 gallons) each day, depending on climate, the

Goats are 'browsers' – gathering branches when confined allows them to follow this instinct.

ambient temperature, the type of diet fed, and the goat's milk yield. A lactating goat requires 1.43ltr (0.3 gallons) of water for every 1kg milk produced.

Water must be kept clean at all times; most goats are fairly fastidious and will not drink from dirty or contaminated water troughs or buckets.

Forage feeding: Most small-scale goat keepers will feed hay as their main source of forage. Late-cut seed hay is generally preferred to soft meadow hay, and hay with a variety of plants such as docks, clovers and vetches is particularly good for goats. However, their natural browsing nature will cause them to pull at hay from racks to select the best bits, and unless suitable racks or nets are used, there can be considerable wastage.

Larger commercial units normally feed silage, and maize silage has proved to be a particularly good forage source, and can be balanced and fed as a TMR (total mixed ration). One problem with silage feeding, however, is the potential for Listeria organisms to multiply in spoiled silage, or in silage left uneaten after 24–48 hours, which should be discarded (*see* Chapter 10).

Maize silage in a clamp on a large commercial goat farm.

Pellets (left) and coarse mix (right), both commonly fed to goats.

Concentrate feeding: The two main choices if concentrate is fed (as opposed to a total mixed ration) are either a pellet or a coarse mix. The latter can be a problem if large numbers of goats are fed together, since early feeders will select the 'best bits', thus potentially leaving poorer quality feed constituents for those who feed later; this is particularly true if there is insufficient feeding space available for all goats to eat together. By feeding a pellet, this problem is avoided.

Never feed more than 40 per cent of the total feed intake as concentrate, or digestive upsets

A total mixed ration fed down a central feed passage.

can occur. Also, remember to introduce new feed constituents gradually to avoid problems.

Deciding What to Feed

- A good starting point is to know the weights of the goats you intend to feed. Ideally try and measure the animal's weight using a weigh crate or similar; with a smaller goat you may be able to stand on the bathroom scales holding it, and then subtract your own bodyweight! Appendix II gives some approximate weights for different breeds and types of goat.
- Use available charts and tables to work out the daily nutrient requirements depending on bodyweight, lactational demands, pregnancy and so on.
- Decide on what feed is available, and is within your budget. Smaller-scale hobby goat keepers are likely to use hay as their main source of forage, supplemented with vegetables, vegetable waste and a compound (coarse mix/cereal/pellet).
- Small groups of goats are often kept at grass during the drier and warmer months of the year, and goats will graze available grass, and also browse and consume other plants/leaves/branches and shrubs. Larger commercial units tend to house goats all year to avoid worm problems.

This goat-handling crate or 'crush' can be adapted to be a weigh-crate.

• Try and assess the bodyweight/condition of all your goats to assess whether or not your diet is being effective. Pet goats have a tendency to be over-fat, and this should be avoided where possible. There are several systems available for assessing the body condition of a goat by handling them.

BODY CONDITION SCORING

Systems are well established for cattle and sheep, with scores attributed in ascending order from 0 to 5, equivalent to a subjective scale from emaciation to grossly over-fat. Such a system relies on the palpation of animals in a standing position, and assessing the degree of fat cover over the ribs in the lumbar area. Unlike the commonly farmed species of cattle and sheep that have a variable covering of subcutaneous fat depending on nutritional intake and production demands, the bulk of the fat stores in the goat are intra-abdominal, thus making a single body condition score based on an assessment of the lumbar spine

An emaciated goat – note the prominent spine and pelvic bones.

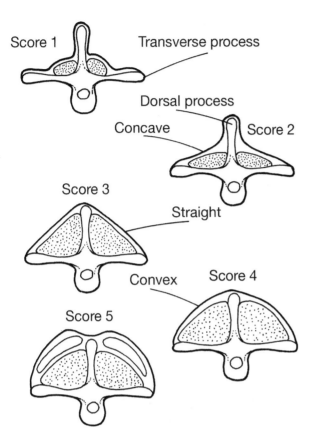

Goat condition scoring – lumbar.

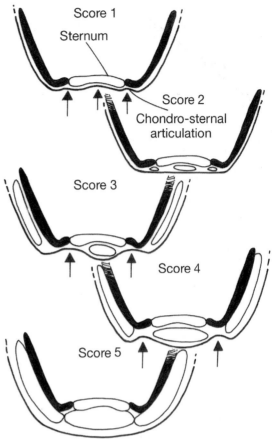

Goat condition scoring – sternal.

less accurate. There is a system that has been adopted in goats, in which the body condition is assessed at two points, namely the lumbar area and the brisket, and an average figure (or dual assessment figure) can then be applied to each goat. This is shown diagrammatically in the accompanying diagrams.

In a commercial herd it would be advisable to condition score all or a proportion of goats at the following times:

- Drying off
- Last two weeks of gestation
- Six weeks into lactation
- Turnout on to pasture (if grazing)
- Beginning of the breeding season

Suggested target condition scores might be as follows:

- Kidding 3–3.5
- Weaning As low as 2, but not below 2
- Service 3–3.5
- During pregnancy 3

GRAZING

One of the most important criteria when establishing a grazing programme for goats is adequate fencing around fields and paddocks, and this is particularly true when grazing near a road to avoid potential accident/injury. It is often said that no fence is ever '100 per cent goat proof', as goats will always attempt to get to that elusive and tempting plant or branch that is tantalizingly out of reach, causing fencing to be bent or broken as a result. Mesh fencing is particularly vulnerable to damage, as goats will stand with their feet on the top; this can be partly offset by running a bar along the top of the fence. Some goat keepers will run electric wires or fencing along the perimeter of a field to protect the fencing.

You might expect 1ha (2.5 acres) of grass to provide sufficient grazing for five to seven goats, depending on grass quality.

Parasite control when goats are at grass is vital – *see* Chapter 7 – and most larger commercial goat farms house their stock all year round because of the difficulty in providing adequate worm control.

Shelter should be provided, but can be temporary in nature. Purpose-made field shelters should be so designed and positioned that they can be easily moved if the ground around them becomes muddy. It may be preferable to have a number of smaller shelters rather than one big one, so that goats at all levels of the hierarchy are protected. Bossy goats may keep more nervous goats from seeking shelter in a single building, particularly if the entrance is narrow.

HOUSING

Most available buildings can be adapted for housing or sheltering goats. The basic requirement is a building that can provide plenty of good air circulation (*see* Chapter 8), plenty of light, a dry bed to lie on, freedom to move around, and good access to food and water. If adapting an old existing building to house a large group of goats, ensure good access by machinery to allow a thorough cleaning and disinfection programme between batches of goats moving through the building, or when needed due to excessive build-up of litter.

As a rough guide, a suggested stocking rate for groups of goats housed together is 1.75sq m (18.8sq ft) per goat. If housed in individual pens (although try and ensure that goats are always kept within sight, sound and smell of companion goats), an area of at least 1.5m by 2.5m (9.9ft by 8.2ft) is required per pen.

A goat being released from fencing in which it had caught its head.

Never house horned and hornless goats together, as the horned ones will dominate and monopolize feeding areas and so on.

Goats are undoubtedly more inquisitive than other species of farm livestock, and unfortunately explore new and unusual things with their mouth, teeth and tongue. Remember also that on their hind legs they can reach to a height of 2m (6.5ft), and younger goats will also readily climb on to feed troughs or window ledges to get better access to what interests them. Pay particular attention to electric cables and switches, water pipes, taps, lagging material, putty, old paint and so on. Putty and old paint (particularly in older buildings, or on items such as old doors or windows used as barriers) can be high in lead, and lead poisoning is a possible sequel. Look at the environment in a building as a goat would, and prevent mishaps occurring.

As with any farm building, try to minimize the risk of injury; you should therefore look out for:

- Projecting sharp objects such as screws and nails that could cause deep puncture wounds, lacerations or eye injuries.

Sharp and damaged galvanized sheets can readily cause injury.

- Broken or torn wire netting; again, lesions can result from sharp, penetrating fragments of wire, and goats will also readily get their heads caught, particularly if horned. This also applies with fencing when they are at grass.
- Galvanized metal sheets can be particularly hazardous when bent or distorted, when the edges can be sharp. Be aware of the danger to goats' feet if bedding rises above the level of the lower edge of the galvanized sheet walls, as goats can inad-

vertently push their foot down the edge of the bedding, and lose a claw when it is pulled back.

TETHERING

This practice should be avoided wherever possible; loose housing or grazing is infinitely more preferable. Tethered goats do require a high level of supervision, with inspections at frequent intervals; they are particularly vulnerable to worrying by dogs and teasing by children. Goats should not be tethered where there are obstacles, and a risk of the chain becoming entangled.

Tethers should be designed and maintained so as not to cause distress or injury to the goats. Collars should be light but substantial, and attached to a strong chain not less than 3m (10ft) in length, with at least two swivels.

Particular care should be taken to provide food, water and shelter, especially during the winter months and in periods of inclement weather, when goats may not have anywhere to shelter. The site of the tether also needs to be carefully thought out; the author has seen a severe outbreak of liver fluke when goats were tethered around the margin of a duck pond (*see* Chapter 7).

CHAPTER THREE

How to Recognize Signs of Ill Health

The background knowledge given in the first two chapters – on the normal goat, its behaviour, structure, function and management – will now enable the reader to begin to recognize signs of ill health, and to decide whether veterinary help is needed.

A good stockperson with a sound knowledge of their goats will develop a sixth sense, thus enabling them to recognize a sick goat by the signs exhibited, many of which can be very subtle. This skill is important, however, because the earlier that illness in a goat is recognized, the more likely that treatment will successfully stem its course.

Many minor illnesses and injuries will heal naturally; others may require some form of medical intervention by the owner (for instance, worms). It follows, however, that the goat keeper should also quickly recognize his own limitations, and know when to call the veterinary surgeon for help. In some diseases goats can deteriorate rapidly, and veterinary attention may still fail to save a very sick goat.

BASIC OBSERVATIONS

- Has the goat suddenly become ill, or has it been getting worse over a few days or weeks?
- Is more than one goat affected?
- What is the age of the goat(s) affected?
- Is it male/female – if a female, is it pregnant?

Contacting your Veterinary Surgeon

When phoning your vet, get a clear idea in your mind of what the main problem is, and also whether it is a real emergency. A goat with a skin problem that is getting worse can wait for a day or so, but a goat with a severely distended abdomen (bloat) will die very quickly and is a genuine emergency. By gathering together some basic information on what is wrong, your veterinary surgeon will be able to decide how urgent the case may be. Have a notebook with you, and jot down your observations, and any other information that will help your vet. Make sure you give clear advice on the exact location of the goat, particularly if it is not at your main address, and give a contact or mobile phone number in case your vet is delayed by another emergency, or gets lost! Find out the map reference co-ordinates; this again may be a useful piece of information.

- Is it still eating, and what is it being fed?
- If milking, has its milk yield reduced or has the milk changed in appearance?

SIGNS OF ILLNESS

The following alphabetical list of symptoms should give the reader a quick indication of what may be wrong, and how severe the problem may be. Further information on specific conditions described can be found in the relevant chapters of this book. However, if you have any doubts or concerns, you must always consult your own vet.

Abdomen – enlargement?

1. The abdomen will naturally enlarge during pregnancy, particularly if the goat is carrying multiple kids; this is quite natural, and should not give cause for concern unless the goat is showing obvious signs of discomfort: this may indicate that kidding is starting, the doe is aborting, or that there are other complications such as a uterine torsion. If you have an accurate service date, then you will know when kidding should occur.
2. Does are quite susceptible to false, or pseudopregnancy (often referred to as 'cloudburst'): the goat will show abdominal enlargement as if pregnant, but the uterus will be full of fluid.

Caseous lymphadenitis abscesses.

3. If the abdomen feels very tight (like a 'drum'), then this is a genuine emergency, and you should seek immediate veterinary help and advice. Bloat is when gas builds up in the rumen, and cannot escape (*see* Chapter 7).
4. Gas can also build up in the abomasum, or in the intestine as the result of a twist or torsion, and resemble bloat. Again this is an emergency, and your vet will need to decide where the gas is building up, so that the correct approach can be adopted to alleviate the clinical signs.
5. If the abdomen has become enlarged over a few weeks and is very firm when palpated, there may be an impaction of either the rumen or abomasum.
6. Ill-thriving kids are often referred to as 'pot-bellied'; there are a number of causes, but the condition is usually related to inadequate early management and feeding.

Abscesses?

These can occur anywhere in the body, but may be visible externally as skin swellings that may or may not be painful, and are related to local infection. Slowly developing abscesses in the head and neck region, although not life-threatening, are nevertheless important, as they may be a sign of caseous lymphadenitis, a potentially serious condition.

Injection-site abscesses can develop if the technique used is unhygienic, but goats do seem to be susceptible to local reactions to vaccines – for example, clostridial vaccines – even if these are administered aseptically.

Anaemia?

This condition can be suspected if on examination the conjunctiva and third eyelid appear pale or even white in colour (examine other goats, if available, for comparison). Possible causes to consider include:

1. Haemonchosis – infestation with *Haemonchus contortus* (barbers pole worm), which can be found in the abomasum and sucks blood.

Injection site abscess.

2. Johne's disease.
3. Liver fluke.
4. Internal haemorrhage following an injury (for instance a ruptured spleen or liver), or after a kidding injury.

Appetite – loss of?

This is a common clinical sign in many illnesses. It may also be an indication of a mouth or tooth problem, or of unpalatable food.

Behaviour – abnormal?

These changes may be subtle, such as being unwilling to enter the milking parlour, or standing away from other goats; or they may be severe, and include circling, toppling over, head pressing or 'star-gazing'. These latter signs indicate a neurological problem, and need urgent veterinary attention (*see* Chapter 10).

Blindness?

Progressive blindness can be difficult to recognize, as an affected goat will learn its environment and will rarely bump into things; however, it often becomes apparent when the goat is moved into unfamiliar surroundings. Sudden blindness is more easily recognized. Causes include:

1. Lead poisoning.
2. Vitamin A deficiency.
3. Damage to the cornea from either a severe kerato-conjunctivitis ('pink eye'), or as a result of damage to the cornea by in-turned eyelids in kids (entropion).

Bloat?

See Abdominal distension.

Breathing – rapid?

Rapid breathing is a sign of a possible pneumonia; check if it is accompanied by a cough, if others are affected, and if the goat is running a high temperature. Veterinary advice should be sought as soon as possible.

Cheek swollen, food dropping from mouth?

This may indicate that there are tooth or gum problems, but may also be a sign of listeriosis, in which the muscles that are normally involved with chewing food and swallowing are partially paralysed.

Collapse/recumbency?

This is characteristic of the terminal stages of a number of diseases, including weakness resulting from excessive loss of weight, in which the goat is too weak to stand. Acute toxic infections such as clostridial disease or mastitis, and neurological problems – for example, listeriosis – will often result in collapse as the condition progresses.

Coughing?

All goats will have the occasional coughing fit, which may be related to food particles 'going down the wrong way'; most of these will resolve on their own, and true choking is rare. A persistent cough, or coughing in a group of goats, gives more cause for concern. Causes include:

1. Infections such as pasteurella (Mannheimia), mycoplasma, lungworm.
2. Allergies such as 'winter cough', thought to be an allergic cough related to the inhalation of fungal spores and dust from hay/straw.

Death – sudden?

Any goat found dead for no apparent reason should give cause for concern, particularly if there are other goats in the same location that may now be vulnerable. It may be worth asking your vet to arrange for a post-mortem examination to be carried out, to establish a cause. These might include:

1. Poisoning (*see* Chapter 16).
2. Acute infections such as clostridial disease, pasteurella.
3. Individual problems such as injuries, heart failure, bloat.

Diarrhoea/scouring?

Causes of diarrhoea in both young kids and adults are many and varied:

1. In very young kids (neonatal), the causes include *E. coli*, salmonella, cryptosporidia (a parasite) and viruses.
2. In older kids, coccidiosis can progressively become a problem, as other neonatal causes become of lesser concern.
3. At grass – worms are a constant problem in grazing goats of all ages, including adults.
4. *Clostridium perfringens* infection is a fairly common cause of severe diarrhoea, often with blood, and followed by rapid death, usually in mature goats.

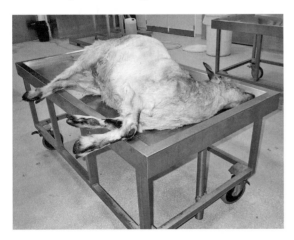

Dead goat at a diagnostic laboratory awaiting post-mortem examination.

5. Dietary factors including sudden diet change, or rapid ingestion of, for example, barley leading to acidosis.

Ear drooping?

Some breeds of goat, for example Anglo-Nubians, have naturally drooping ears, but the carriage of the ear can be abnormal for a variety of reasons, including the goat itself just being 'off colour'. Other more specific causes include listeriosis due to nerve damage, middle ear infection, or as a result of the development of an aural haematoma (blood blister), often related to head shaking due to ear parasites.

Ear swollen?

See 'Ear drooping'.

Eye blindness?

See 'Blindness'.

Eye running?

This is usually an indication of conjunctivitis, with excessive tear production. In young kids it may be related to entropion (turned-in eyelids, so that the eyelashes rub on the cornea); in older goats it may be associated with an infectious keratoconjunctivitis ('pink eye'), and usually more than one goat will be affected.

Fits/convulsions?

This condition may be related to a neurological problem, or a metabolic upset, for example pregnancy toxaemia or poisoning. It needs urgent investigation.

Head shaking?

Could be linked to ear mites or fly worry; excessive head shaking can lead to the development of an aural haematoma (*see* Ear drooping?)

Itching/hair loss?

This condition is most likely to be related to mange or lice infestation, particularly if more than one goat is affected. Pygmy goats can develop a specific condition referred to as the 'pygmy goat syndrome'. Consult your veterinary surgeon, as there are many causes (*see* Chapter 12): once the cause has been determined, then the correct treatment can be applied.

Jaw swollen?

If the bony part of the jaw is swollen it may indicate a tooth root abscess. Another possibility is a fractured jaw, particularly if there has been a history of fighting. Soft 'pitting' swelling under the jaw (so-called 'bottle jaw') can be an indication of protein loss or anaemia, as seen in liver fluke infestation or a worm infection involving haemonchus.

Lameness?

This is a relatively common problem, and needs a full examination (with veterinary consultation if necessary) to establish a cause. There can be varying degrees of severity, depending on whether the goat is lame on one, two or more limbs, and on the type of lameness itself. The limb may be held at a bizarre angle if there is a fracture, for example, with no weight-bearing on the affected limb. Most lameness is related to foot abnormalities.

Lips – scabby?

Lesions that are seen most commonly are related to infection with the orf virus. Hair loss

Lame goat – note the way the affected limb is held up.

and scab formation can occasionally be seen in bucket-fed kids, when the cream/fat from the milk may be deposited over the muzzle.

Poor growth rate?

A condition often reported in adult goats is the wasting or fading goat syndrome. There is no single cause, and a full investigation is normally required in conjunction with your veterinary surgeon to identify a cause. Is it an isolated occurrence in just one goat, with others in the same group performing well, or are others affected? Causes are many and varied, and include parasites, Johne's disease, poor nutrition, mineral deficiency.

Scouring/diarrhoea?

See Diarrhoea/scouring.

Straining (tenesmus)?

As straining can be an indication of a problem in the digestive, urinary or reproductive tract, a full clinical examination with input from your veterinary surgeon is advised. Possible causes include:

1. Severe enteritis, particularly when the rectum becomes thickened and inflamed – this can occur in, for example, coccidiosis.
2. In breeding does, it may be an indication of parturition/dystocia, or of an imminent prolapse of the vagina or cervix.
3. In older does, straining with the passage of a foul-smelling discharge can indicate the presence of a tumour in the reproductive tract.
4. In males, the bladder can occasionally become blocked with a small stone or calculus, causing intense discomfort, crying and constant straining – a genuine emergency.

Swaying/staggering gait?

In young newborn kids, this may be an indication of congenital swayback or of border disease, both resulting from damage to the nervous system of the developing foetus in the uterus before birth, and related to copper deficiency and a viral infection respectively. Swayback can also occur in a delayed form in older kids. Spinal abscesses can cause similar signs.

Thinness?
See Poor growth rate.

Udder swollen?
This is a sign of mastitis, which although not common in goats, can be very severe with gangrene developing. In milder cases, the only other sign may be a few clots in the milk.

CHAPTER FOUR

'Prevention is Better than Cure'

Although modern goat keepers have many sophisticated vaccines, antibiotics and other pharmaceutical products at their disposal, the maintenance of a healthy herd is still predominantly down to good stockmanship. We know that some people are very good with animals and make good stock-keepers, often having a 'sixth sense' that things may be wrong. However, all goat owners must work at developing their stock skills – remember, your goats rely on you for their well-being – and if you let them down, they are more likely to become ill or fail to thrive. A skilled goat keeper will:

- Know his goats, and even in large groups will recognize early signs of ill health in individuals, often purely by subtle changes in behaviour.
- Anticipate problems before they happen, and take steps to avoid them; examples include worms at grass in summer, pneumonia when housed in winter, and overgrown feet that may need trimming.
- More specifically, take steps to carry out recognized routine control measures such as regular clostridial vaccination.
- Carry out regular observation of all goats at least daily, and if unable to do so, make sure that a competent person checks them on his/her behalf.
- Act promptly and responsibly if a sick or injured goat is identified, seeking veterinary help if necessary.

In the past few years there has been a noticeable trend towards the establishment of protocols, so-called 'health plans', to ensure that farm livestock health and welfare is maintained. Although farmers and their veterinary surgeons have always worked together to achieve this, it has often been fairly unstructured and haphazard.

Health plans recognize that every farm or livestock unit is different, therefore measures to control disease should also recognize this. These principles apply equally to a goat unit with 2000 goats, as to one with a single pet goat. Quite obviously there are major differ-

This goat is suffering from pneumonia; note how it has separated itself from the group, and stands with its head extended.

30

ences between these two units, but the owners of each will have the same ultimate aim, namely to ensure that the goat(s) for which they are responsible remain fit and healthy (and on the larger unit remain productive).

A health plan can be a single sheet of paper, detailing all the procedures that need to be adopted, but for larger units it is likely to run into several pages. Many veterinary surgeons are familiar with the concepts involved, and will be pleased to help you establish such a protocol.

KEEPING RECORDS

Although it is undoubtedly a chore, it is important to develop a recording system for your goats. The system you adopt must be user friendly, and produce information that can be readily retrieved and is useful and relevant, such as planning your future breeding and replacement programme, or deciding on which goats need culling. There are many options available, from the simple to the sophisticated:

- A dedicated diary is entirely adequate for a few goats.

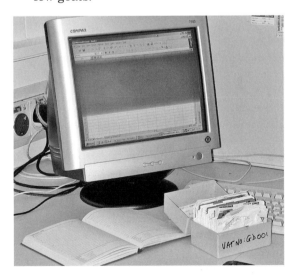

There are many ways of recording information on your goats' management/health.

- A diary can also be useful to record major events (such as feeding changes, and other management procedures such as turn-out, vaccinations and so on) on larger units.
- On larger units, however, it is worthwhile adopting a system that can provide individual health records for mature animals, including, for example:
  - milk yields
  - service dates
  - kidding dates
  - kidding difficulties
  - kids born
  - prolapses
  - vaccination history
  - illnesses
- Such a system could be, for example, a file-card system, or it could be computer-based.

Movement Records

Remember also that it is a statutory requirement in EU countries to maintain a movement record, and this must be kept up to date. In the UK, Defra, through the network of local Animal Health Offices, will be able to provide guidance, but an example of the type of information to be recorded is given on p. 32.

You must record all movements on and off your holding, including movements to, for instance, shows, markets, rented land, or to a veterinary surgeon for treatment, within thirty-six hours of the move taking place. Although onerous, such information becomes vitally important during outbreaks of notifiable disease in the country, such as foot and mouth disease, when it may be necessary to trace the movements of susceptible animals quickly, particularly if disease is related to a gathering such as a show or market.

All existing goat units should already have been registered, and this is an EU requirement; again, new goat owners in the UK can get more information from Defra. In the UK each unit is issued with a number referred to as the 'holding number', or CPHH number (County/Parish/Holding/Herd), and this should be quoted in all official correspon-

An official Defra record used for sheep and goat flock/herd movements.

dence. A number of helpful and informative leaflets are available, including the *UK Goat Welfare Codes*, which are a valuable source of information.

Your movement record should include:

- date of movement
- number of goats moved
- identification mark (individual/herd prefix, depending on the numbers involved)
- address of previous holding if not your premises (e.g. market/showground)
- destination address
- if moved from a market, the lot number
- if sold without leaving the holding, the name and address of the new owner
- replacement tags: cross-reference with previous marks

Goat Identification

For any recording system to be effective, there must be a means of accurately identifying your goats. There are many ways of achieving this, such as eartags, collars and tattoos, and the inexperienced new owner may assume that pet goats do not need any identification. It is a legal requirement, however, for all goats to be identified, and new EU legislation requires all goats to be individually identified with an eartag. Goats may not tolerate eartags as well as sheep or cattle, and local reactions may develop, tags may be lost, or ears may be torn. Seek advice on the most suitable tag to use, and where to place it. Tags available are many and varied, and examples are given below.

If you have pet goats or valuable pedigree animals, it may also be worth taking photographs in case they escape or are stolen.

BIOSECURITY

The greatest risk to the health of a goat is being in the company of another goat! It is important to bear in mind, however, that there are a number of infectious agents that can spread readily between sheep and goats (for example, caseous lymphadenitis) or cattle and goats (for example, Johne's disease).

A pet goat kept on its own away from other goats (and sheep/cattle), is unlikely to develop an infectious disease. If the owner decides to buy a companion, however, then there is a two-way risk of infectious disease developing:

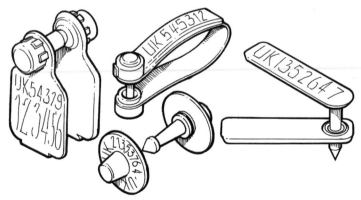

There are many types and makes of eartag available for goat owners.

1. The incoming goat may be carrying or harbouring an infection that the established goat has never encountered, thus potentially resulting in disease developing in the group of goats into which it is moved.
2. Conversely, however, it may well be that the existing goat(s) – that is, the goat(s) into whose company the purchased animal is to be moved – is carrying or harbouring an infection to which it has already produced antibodies, and is therefore immune. But the incoming goat may be susceptible, and on this occasion it will be the purchased goat that will become ill.

This same principle can be applied to larger herds of goat, and we use the term 'closed herd' for one that has not purchased any goats for a number of years, and thus has a pool of infectious agents that all goats will have encountered, and to which they will have developed immunity (although there can be exceptions, particularly if youngstock are reared away from the home unit and then re-introduced). A truly closed herd is one that rears its own replacement goats, and never purchases any 'outside' stock. If the 'occasional' animal, such as a replacement male, is purchased, or some of the goats visit shows on an occasional or regular basis, the herd is not, strictly speaking, closed! Unfortunately, most of the larger units have to buy in a male now and then in order to improve genetic status, if they are not using artificial insemination.

To maintain their closed herd status, it is important that the unit has secure housing and fencing to prevent stray goats inadvertently bringing infection on to the unit, or for goats on the closed unit to escape to a neighbouring farm. If at grass, ensure that your goats cannot have nose-to-nose contact with goats or sheep in a neighbouring field. Gateways are particular problem areas, as are single-strand wire fences – hedges afford more protection. Consider 'double fencing' these areas, so that you have two fences or gates separated by a gap of at least 3m (10ft).

Although an important part of hobby goat keeping, there is nevertheless a risk of disease transfer at shows.

Guidelines for the Purchase of New Goats

If you have to buy in replacement animals, for whatever reason, then there are a number of guidelines designed to help you buy in only the goats themselves, and not the bugs that may accompany them!

- Try to buy direct from another farm (preferably animals bred on that farm), and try and limit your purchases to as few farms as possible. Although not a common practice with goats, there are many cattle and sheep purchased through markets or dealers, making it much more likely that disease will be introduced due to the mixing of animals from a number of differing sources prior to sale.
- Is the farm a member of any recognized health scheme that would confer freedom from certain diseases, such as caprine arthritis encephalitis (CAE)?
- Always record in your movement book the exact origin of every goat purchased (a statutory requirement, but also useful if problems arise that can be avoided in future!).
- Consider the part of the country that the goats are coming from, and how they have been kept. If they have been grazed in the wetter western areas, could they bring

33

liver fluke on to your farm? If they have been housed throughout their lives, they will never have been exposed to worms (an important consideration if you intend to graze them outdoors).

- Ask about the health status of the goats to be purchased: your veterinary surgeon should be able to advise. He/she may even know the goats you are purchasing, or be able to advise you of suitable goats with his/her local knowledge. He/she may be able to discuss the health status of goats you are buying with the veterinary surgeon that looks after the farm you are buying them from.
- Discuss with your veterinary surgeon whether any laboratory tests may be useful, and carry them out pre-purchase. Examples may include CAE or enzootic abortion. Again, your veterinary surgeon may be able to help arrange these tests before the goats arrive at your farm, in liaison with the vendor's veterinary surgeon.
- Ask to see a full history of all recent preventative measures undertaken on the vendor's farm, such as vaccinations, worming doses and so on. If these are not complete, then ensure that you carry out any necessary procedures whilst the goat

is in quarantine. Consider using a quarantine worming dose regime – for example, a simultaneous dose of a levamisole and an avermectin product (*see* Chapter 7), to avoid bringing resistant worms on to your farm. Your vet will be able to explain this principle to you.

- I would advise against purchasing goats that have been vaccinated against orf, Johne's disease or enzootic abortion, because it may well be that the unit you are buying them from has a history of disease (hence the reason for vaccinating), and with these conditions, vaccination will not guarantee freedom from disease, and you may inadvertently be introducing infection.
- Try and use your own transport if possible, making sure the vehicle has been cleaned and disinfected before leaving your premises; this procedure should be repeated when you return.
- If you are purchasing goats from abroad, be fully aware of all important legislation requirements to prevent the import of exotic diseases not seen in the UK (in addition to the more common diseases already discussed). Remember also that they may be totally susceptible to a disease on your farm that they have never encountered in

Veterinary laboratories are able to undertake a wide range of laboratory tests to demonstrate exposure to, or freedom from, many diseases.

It is important that vehicles travelling between livestock premises are kept clean.

their country of origin. For further information on importing goats, talk either to your local government (Defra in the UK) Animal Health office, or your own veterinary surgeon.

- Be prepared to walk away from a purchase if problems or obstructions are encountered at any stage of the buying process. Avoid buying the 'bargain goat', or one that looks like a 'waif wanting a home' if you want to avoid bringing problems back to your premises!

Quarantine Arrangements

It is important that you have a facility to allow new goats to be kept in quarantine, before being mixed with your own goats. There is much debate over what is a suitable period of time for a goat to remain in quarantine, and for most diseases two to three weeks would be acceptable. Ideally, however, aim for at least six weeks, and even in this timescale certain diseases, such as caseous lymphadenitis (CLA), may not have sufficient time to become apparent if the goat were incubating disease on arrival. A quarantine facility should be:

- Exactly what it states! It must be a part of the farm/smallholding where goats can be kept separate from other goats (and sheep, as some infectious diseases are common to both). If it is a building, it must be a separate structure situated at least 6ft (2m) away from any other building housing goats or sheep. A small paddock will suffice, but again it needs to be completely separated from goats and sheep in adjoining fields. Any gates or single strand fences where nose-to-nose contact could occur should be double fenced with a 3m (10ft) fence separation.
- If the quarantine facility is a building, then it should be capable of being thoroughly cleaned and disinfected after the period of quarantine is complete.
- It should allow easy access for regular observation of quarantined goats, with good natural or artificial lighting.

Your veterinary surgeon will be able to help and advise on the suitability of your chosen quarantine area. Once the goats are in quarantine, ensure that you observe them daily, making time to handle each one at least weekly for signs of CLA (lymph node enlargement), and also to examine their feet for, for instance, footrot.

Ideally, those handling/feeding the goats in quarantine should not immediately handle other stock, to minimize the risk of spreading disease. Have a dedicated pair of overalls to wear when examining quarantined animals, and keep these overalls in or near the quarantine area. Dip your boots in a disinfectant footbath before visiting other goats. This may all appear to be excessively fussy, but careful thought and keeping to basic hygiene procedures may prevent a major outbreak of disease.

Consider, in consultation with your vet, if prophylactic (preventative) doses of, for instance, a wormer or antibiotic may be beneficial before quarantined animals join the existing goats.

Isolation Facilities
All livestock units should also have isolation facilities in which sick goats can be placed,

A disinfectant footbath is always useful to have outside your quarantine facility, or outside any pen in which you have sick or ailing goats.

if they are affected with a disease that may spread rapidly, such as salmonella, or infectious causes of abortion such as enzootic abortion. Your quarantine facility can 'double up' as an isolation facility, but it must be readily capable of being cleaned and disinfected.

Other Biosecurity Measures

If you are familiar with local pig and poultry units, you will be aware just how seriously they take the risk of disease being introduced by visitors, lorries and suchlike.

I have visited pig units where I have had to remove my day clothes, have a full shower, before donning disposable underwear and overalls before I am even allowed to see the pigs. If visiting a number of units in the same ownership, then this procedure (shower in/shower out) has to be repeated on each occasion – and I end up being very clean!

I would not advocate this approach on any goat unit, but there are still a number of recommendations I would make to protect your goats, and these measures were strongly endorsed after the foot and mouth epidemic of 2001, where there was clear evidence of spread of infection between farms by a variety of means.

- Ideally, discourage members of the public (particularly those with their own goats) from having direct contact with your goats. It is appreciated that part of the fun of keeping a goat is sharing views and ideas by visiting each other's premises, but do bear in mind the disease risks involved, and take sensible measures to avoid spreading infection. Shows pose a particular problem, when it is only natural for other goat owners and visitors to touch and stroke those on show.
- Try and encourage visitors to park vehicles away from the goat buildings, particularly in wet weather. Any manure/mud carried from one farm can easily drop off when the vehicle is parked. If you are concerned that your own vehicle may be contaminated after visiting a farm or market, then wash

it down as you leave, or when you return home, and disinfect the wheels and the wheel arches.
- Every goat unit will suffer the occasional death, and if you need to arrange a post-mortem examination, ask your veterinary surgeon. Your vet may undertake the post mortem, or he/she may arrange for the goat to go to a veterinary diagnostic laboratory. Whatever happens, speed is of the essence: do not leave dead goats lying around as they will quickly decompose, and may also be attacked by predators such as foxes that may even drag a smaller animal away, thus potentially spreading disease around. If a dead goat is to be collected for disposal, then try and place it in a solid container for collection away from the buildings, so that the collection vehicle (which may have visited other farms) does not have to come right into your premises.
- Increase the frequency of removing faeces from buildings, and cleaning. This will reduce the chance of faecal build-up in the building, and potential run-off from one part of the building to another if, for example, water troughs or guttering leaks. Always clean buildings thoroughly between batches of goats (for example, during kid rearing).
- Adopt an 'all in, all out' system where possible, and avoid the temptation to hold

These dead goats have been left lying around for eventual collection by a disposal agent.

36

Disinfectants

Ensure that you have an approved disinfectant available to use at all times. Read the instructions carefully: these will tell you what the disinfectant is effective against, and at what concentration it should be used. It is important to get the concentration right, or it may not be fully effective. Disinfectants will not work as well in the presence of organic material such as faeces, so try and ensure that any surfaces to be disinfected (wheels, boots, walls or floors) are visually clean before applying the disinfectant solution. Equally, if the disinfectant in a footbath becomes contaminated with faeces, it will quickly become ineffective; therefore change the bath frequently.

Be careful about disposing of unwanted disinfectant, as it can be harmful to aquatic life – for example, fish – if poured directly into a stream.

---

back from a batch even one goat that hasn't thrived to join the next batch. It will carry all the micro-organisms that were active in that batch into what may well be a clean batch.

- Avoid spreading farm manure (any species) on grazing land where possible, unless at least three months has elapsed, and faecal material is no longer visible.
- Keep the farm tidy – this will help discourage vermin. If rats and mice are a problem, consider seeking the help of a local pest controller if you are unable to control them yourself. Rats in particular can harbour and transmit a number of infectious organisms to which goats are susceptible, including salmonella, yersinia and leptospirosis.
- Ensure that feed stores are bird- and vermin-proof; birds can carry organisms such as salmonella in faeces. Place open bags of food in bins so they cannot be contaminated with rat or mouse droppings.
- Check hay and straw bales regularly for rats' nests, and destroy any that are visually contaminated. Also discourage cats

from sleeping in hay or straw barns, and discard any bales on which cats may have produced a litter of kittens – cats are an important part of the toxoplasma life cycle, an important cause of abortion in goats.
- If borrowing farm equipment such as a weigh crate, ensure that it is free of faecal contamination and is fully disinfected before use.

THE SAFE USE AND STORAGE OF MEDICINES

There are many types of 'medicine' that you may need to administer to your goat at some stage in its life, such as a vaccine dose or wormer. However, the array of products available, and the widespread claims that are sometimes made by the manufacturers, may be confusing. Confidence will come from experience, and the novice would be well advised to learn from other owners competent in the procedures involved. They may be able to enrol on courses at local agricultural colleges that are directed towards sheep owners, but that could be equally beneficial for those keeping goats. A further option is to ask your veterinary surgeon to show you some basic techniques, and assess your ability to carry it out effectively and safely; such information will be readily given, particularly if the product is purchased from the vet!

Simple golden rules to follow include:

1. Read the label, and the insert in the package, if available.
2. Check you are giving the correct dose.
3. To do this accurately, you need to know the weight of the goat you are dosing. A goat's weight can be difficult to assess accurately, and many goat owners underestimate the weight, and underdose as a result (*see* Chapter 2 and Appendix II). If you are dosing a number of goats, always dose to the heaviest in the group: if you dose to the average weight, then the heaviest will be underdosed.

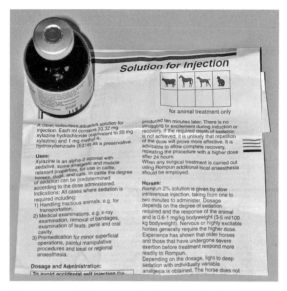

Keep any veterinary product you use in its original packaging where possible, and retain the insert.

4. Check that you are giving the product by the correct route.
5. Is the product licensed for use in goats? This is a big problem in many countries due to the lack of products with a full licence. Alternative unlicensed products can be used so, if in doubt, consult your veterinary surgeon.
6. Remember that many products will have a declared withholding time for milk or meat and, if they do not, you may have to apply standard withholding times (typically 7 days for milk and 14/28 days for meat).

Oral Administration (by Mouth)

Drenches

The medicines most commonly given this way are mainly wormers in liquid form. Be aware that certain wormers have the same name regardless of whether they are given by mouth or by injection, and it is vitally important that you check the correct route of administration by reading the label.

Other medicines given by mouth include vitamin and mineral preparations, and other nutritional supplements.

If you have only a small number of goats to treat it is probably not worth buying a drenching gun; using old syringes that have been cleaned out is a useful alternative method. Having worked out the dose of drench to be given (usually from the weight of the goat), fill either your syringe or drenching gun with the correct dose. A drenching gun is particularly useful because it can be preset to give exactly the same dose to a number of goats, which will save time having to measure each one out individually. If you are using a drenching gun be particularly careful to ensure that it is maintained correctly and kept in good working order. The shape and design of the nozzle is important, and if this is bent or damaged you may cause injury to the soft tissue around the throat whilst dosing.

When drenching goats hold them gently but firmly. The head should be held securely with one hand cupped under the lower jaw, and then gently tilted up a little; insert the nozzle of the gun into the mouth on to the back of the tongue, and give the dose by pressing the plunger. If you are using a syringe, this can be pushed into the corner of the mouth (again while the head is held slightly up), and the contents squirted out. Be careful not to get it between the goat's back teeth, as it will end up in splinters!

Boluses/Tablets

Some medicines come in solid form as either boluses or capsules, and these should be administered using a gun supplied by the manufacturer. This is usually a long, hard, plastic tube with a plunger, the bolus being placed at the end of the tube that is then gently pushed to the back of the mouth and the plunger depressed. With experience you will be able to tell if the goat has swallowed the bolus, but keep a careful watch for a few minutes in case it is spat out on the floor.

If a tablet has to be given without the aid of a gun, it is most important to protect your fingers, as a goat's teeth can be very sharp and cause severe damage. The safest approach is to put your thumb in the front of the mouth (there are no top teeth), and press up on the dental pad, and the mouth should then open. The tablet can now be placed on the back of the tongue, and the mouth held closed – and then you just hope that the animal swallows the tablet!

Injections

Many medicines, including most vaccines and antibiotics, are given by injection. Watch out for training courses at local agricultural colleges (usually for sheep or cattle), and try and attend a training course before you attempt to inject your own goats. Alternatively, ask your vet to demonstrate the procedure.

There are several different routes for injections, and it is important that you read the instructions to know which route you should be using for any particular medicine. Also make sure that you are using the correct size of syringe and needle. Needles are measured by gauge (g): the higher the gauge, the thinner the needle. Always use clean needles, and if you drop one, discard it.

How to Fill a Hypodermic Syringe

1. Shake the bottle well, particularly if the contents are in suspension.
2. Insert a sterile needle into the bottle, attach a syringe and turn the bottle upside down.
3. Pull down the plunger to the required mark.
4. Withdraw the needle still attached to the syringe.
5. Cover the needle with its guard.
6. Hold the syringe vertically with the needle upwards.
7. Gently tap the syringe to move any air bubbles to the top.
8. Depress the plunger to remove the air from the syringe.
9. You are now ready to inject.

Intramuscular Injection

This is an injection given into a muscle mass; the most common sites are the neck muscles and

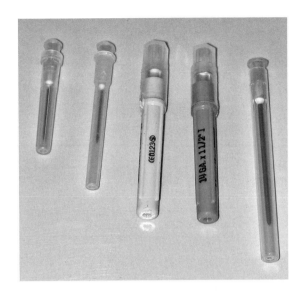

There are many different sizes and lengths of needle available; make sure you use the correct one.

the hind-limb muscles. Be aware that there are some large nerves running through the muscle masses in the hind limbs, and it is advisable to inject into the muscle at the front of the limb rather than the muscle at the back of the limb where you may hit the sciatic nerve.

Subcutaneous Injection

These are injections given under the skin, and are commonly used in goats for vaccines and injectable wormers. The two main sites used for injection are either on the side of the neck, or on the side of the chest. Under certain circumstances, vaccine can also be given in the subcutaneous tissue over the lower front part of the chest.

The skin should be lifted into a 'tent', and the needle inserted into the space between the two edges of skin. Be careful not to push the needle right through the other side.

Again, problems can develop from injection site abscesses if either dirty needles are used, or the injection is made through dirty skin. Remember also that some products (particularly vaccines) can result in a localized abscess, which appears to be a reaction to the adjuvant (vehicle) in which the vaccine is suspended.

Needle Sizes

For kids, a 21g, 5/8in needle (green cover) is suitable, but as the muscle masses in very young kids are extremely small, it is probably best to avoid intramuscular injections, and to use oral medication or subcutaneous injection instead.

For adult goats, a 19g, 1in needle (yellow cover) is usually adequate, though if the injection material is very thick you may need a larger needle; your vet will advise.

These needle sizes can be used for both intravenous and subcutaneous injections.

Be warned that if you use dirty needles, or inject through a dirty skin, then a subcutaneous injection abscess can develop.

Intravenous Injection

This procedure involves giving a product directly into a vein (using the jugular vein); it is a skilled job, and should only be undertaken by your veterinary surgeon (or in consultation with your veterinary surgeon).

MEDICINES' LEGISLATION AND SAFETY

Goat owners must be aware of their responsibilities for the safe and responsible storage, use and disposal of medicines. Details of all medicines used, including alternative therapies (for example, homoeopathic remedies), must be recorded at the time of use. These records must include all treatments and medicines administered, including those administered by a veterinary surgeon. The following information must be recorded:

- identity of medicine or therapy;
- quantity of medicine or therapy;
- date of purchase;
- date of administration;
- name and address of supplier;

- identification of the animal or group of animals to which administered;
- the number of animals treated;
- dates when meat and milk become fit for human consumption;
- name of person administering the medicine or therapy.

It is strongly recommended that the following details are also recorded:

- length of withdrawal periods for milk and meat (if applicable);
- batch numbers of medicines used;
- expiry dates.

It is also recommended that the following 'code of practice' be observed:

- All medicine records must be retained for a period of three years. Records relating to the use of 'prescription only medicines' (POM) must be retained for five years.
- All medicines must be stored securely under lock and key. It is recommended that the storage be separate from the milking parlour and milk storage area. Only medicines for immediate use should be available in the parlour.
- All medicines must be properly labelled in accordance with the legislative requirements, and used and stored according to the instructions.
- The prescribing veterinary surgeon must inform the administrator or animal keeper of the appropriate withdrawal periods, and all withdrawal periods must be observed.
- Only authorized medicines, or those under the specific direction of a veterinary surgeon, must be used.
- Sharps (for example, needles) and any unused medicines must be disposed of safely and in accordance with instructions from the supplier.

CHAPTER FIVE

Reproduction and
Reproductive Problems

Goats have a seasonal breeding cycle, and in the northern hemisphere the breeding season extends from September to March; between March and August the majority of females are in anoestrous (not cycling). There can be some individual variation, however, when a goat occasionally breeds outside these months; the Anglo-Nubian and Pygmy breeds show the greatest propensity for this.

The stimulus for the commencement of the breeding season is decreasing day-length, which applies equally to males and to females. Other factors having an influence on the female breeding cycle include temperature, body condition and underlying nutrition, and the presence of the male.

This seasonal breeding cycle created early problems on commercial goat milking units, because they were unable to supply milk throughout the year, as demanded by supermarkets and consumers. Out-of-season breeding by the manipulation of day-length has now eliminated this problem (*see* later).

BASIC REPRODUCTIVE DATA

Breeding season: September–March (N. hemisphere)
Puberty: 5 months
Age at first service: 4–6 months (male); 7–18 months (female)

Oestrous cycle: 19–21 days (some breed variation)
Duration of oestrous: 24–96 hours (average 36–40 hours)
Ovulation: 24–48 hours after start of oestrous
Gestation length: 150 days (range 145–156 days)

OESTROUS

When kept with other goats, and particularly males, a goat on heat will show a number of fairly characteristic signs including:

* tail wagging;
* frequent bleating – this can occasionally be very loud and persistent, and can be mistaken for signs of pain or discomfort!
* dribbling urine;
* swelling of the vulva with a slimy mucous discharge occasionally showing;
* a slight rise in milk production, which may occur 8–12 hours before the start of oestrous, then drop below normal during oestrous itself.

When kept singly, however, oestrous may not be so obvious and can be easily missed. Some goat owners use a 'billy rag' (a cloth rubbed over the head and genital area of a male to capture its scent) which, when placed near a doe suspected to be on heat, will often cause her to show more typical signs of heat that can be more easily recognized. The billy rag is usually kept sealed in a 'billy jar' to contain the somewhat strong and unpleasant odour!

Unlike cows, female goats will not normally mount and ride each other when on heat.

A buck running with females will readily recognize a doe in oestrous, which is the only time she will allow normal mating activity to occur. If taking the doe to a buck for service (or vice versa), or if using artificial insemination (AI), then recognition of this true oestrous becomes vitally important if a successful pregnancy is to follow.

The reader is advised to separate males from females at weaning, to avoid the possibility of an unwanted pregnancy – bucks can rapidly become sexually active.

It is important that the reader has a basic understanding of the hormonal changes that take place during pregnancy, and these are detailed in the diagram below. This shows how a follicle ruptures, releasing an egg that in turn passes through the oviduct for fertilization and implantation in the womb. More than one follicle may mature in each ovary, thus leading to a number of eggs being potentially fertilized. The follicle is replaced by a developing corpus luteum (CL) that will begin producing progesterone, and this CL will be maintained through pregnancy. If fertilization does not occur, another hormone known as luteinizing hormone (produced by the pituitory gland) causes the CL to regress, with a subsequent reduction in circulating progesterone. This in turn results in a wave of follicles developing, with one or more becoming dominant; and a gradually increasing level of oestrogens results in the behavioural changes seen in oestrous. The cycle is then repeated.

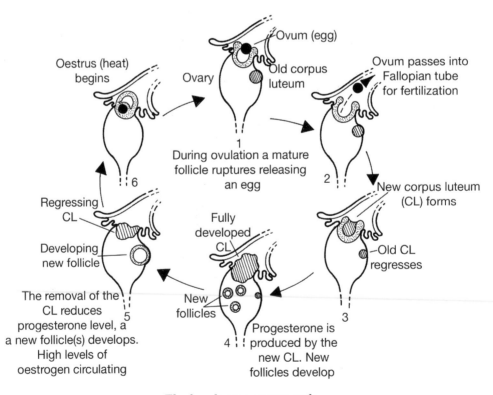

The female goat oestrus cycle.

THE BUCK

It is hard to justify keeping a male goat if you have only a few females. They are strong and very smelly, and many goat keepers may not feel able to provide them with the necessary level of care, and to keep them safely without trouble. Also remember that with only a small number of goats, it can be difficult to avoid in-breeding (when a male mates with his own daughters or grand-daughters for example), and this can result in many undesirable characteristics. The owner with only a small number of goats will probably resort to finding the nearest available male (remembering the importance of biosecurity, *see* Chapter 4), or consider using AI. New goat keepers should join a local goat club, where many of these problems can be discussed and solutions found!

The selection of a good breeding male is vital not only for a successful pregnancy, but also to ensure that the offspring are viable and productive (in commercial milk and meat enterprises). The science of sire selection is less well defined in goats than other farm species, but in simple terms, male offspring of high milk-producing does are likely in turn to sire high milk-yielding offspring, and similar principles will apply to fibre-producing, or meat goats.

Bucks fighting.

If a buck is to be purchased, try and purchase one from a known reputable pedigree source. Remember that not only must he be of good conformation and fertility, but he must also be healthy – there are many reports of males (all farm species) being purchased and brought on to farms, only to introduce an infectious disease because he has not been examined, subjected to laboratory tests, vaccinated, wormed and placed in quarantine.

If you are investing money in the purchase of a good buck, then it may well be worth consulting your veterinary surgeon to develop a pre-purchase health package, paying particular attention to the hind-limb conformation and feet (avoid bucks with badly overgrown feet, footrot or scald). Make sure he is placed in quarantine before joining the other goats.

Examine the testicles and penis. The testicles should be of an even size, and feel firm and 'springy'; they should not be excessively hard or flabby, and should move freely within the scrotum; and there should be no signs of pain or discomfort whilst the scrotum is being handled. Do not purchase a buck with only one testicle (no matter how good his pedigree is!). There should be no abnormal swelling above the testicles; this may indicate the presence of an inguinal hernia (which can follow fighting among young bucks). The penis should be examined whilst still in the prepuce initially, and if there is any evidence of pain on handling, of abnormal swelling, or of a discharge, then this may indicate that all is not well.

If rearing your own replacement bucks, remember that the time of onset of puberty is influenced by increasing bodyweight and by the time of year. Thus male kids born in early spring will inevitably reach maturity in their first autumn, whereas late-born, small kids that do not grow well, may not. Libido increases from late September/early October (northern hemisphere), reaching a maximum in December, and declining gradually to reach a minimum in July, then increasing again during August, thus repeating the cycle.

In theory, one buck can mate up to 100 does during one breeding season, but a ratio of one

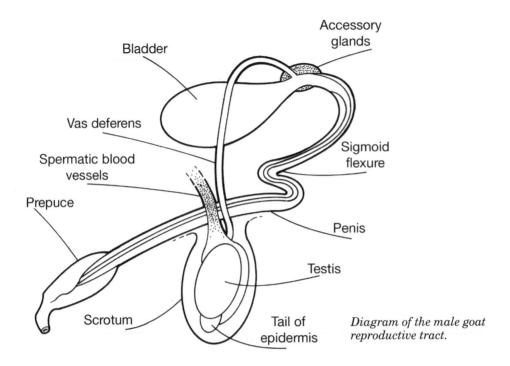

Diagram of the male goat reproductive tract.

mature buck (or two young bucks) to 50–70 does is probably a more realistic ratio. This ratio may have to be reduced dramatically, however, perhaps to as low as 1:5, if does have been synchronized artificially to bring them all on heat together.

ARTIFICIAL BREEDING

Out-of-season breeding is becoming widely used to enable commercial producers to maintain all-year-round milk production, and for fibre producers to benefit from three kid crops in two years.

The principle is to keep goats (males and females) under an artificially long daylight regime during the winter months, followed by a sudden change to normal day length in the spring, thus enabling out-of-season breeding to be achieved in the northern hemisphere from April to June, during the normal period of anoestrous.

From 1 January, provide twenty hours of artificial light for sixty days (this can be achieved by using a significant number of fluo-

---

Care of the Buck

The buck is often the forgotten member of the herd, except during the breeding season. Make sure you include him in all routine procedures undertaken in the remainder of your goats; ensure that his feet do not get overgrown, and that he is vaccinated (and boosted) against, for example, clostridial disease and so on.

Prepare the buck nutritionally for the start of the breeding season as you would the does. He needs to be in good condition (but not overfat), and his fertility will often be improved by having him on a rising plane of nutrition leading up to his first service; this should ensure maximum fertility.

Bucks can become very 'lonely and isolated' if kept singly away from other goats. Try where possible to have a companion goat (an uncastrated male), or at least keep him within the sight and sound of other goats.

---

rescent tube strip-lights for example). After sixty days, return the goats back to normal lighting, when oestrous should follow seven to ten weeks later. The oestrous period may

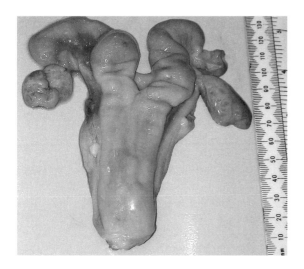

The female genital tract.

be shorter than normal, and be less obvious, so best results are achieved by running bucks with does, with the sudden introduction of males after the return to normal day length further increasing the percentage of successful matings.

On commercial units, another approach often adopted is to synchronize goats so that they can be served in groups, thus enabling milk yields to be manipulated to meet market demands, and also providing a useful management procedure to plan for periods of kidding activity. Such an approach may utilize vasectomized bucks, or progestagen-impregnated intravaginal sponges or other hormonal regimes.

INFERTILITY

Failure to breed can be a disappointment to the 'hobby goat keeper', but a disaster for the commercial goat farmer. There are many reasons for an apparent failure to breed, and unless the problem is easily apparent, it is probably advisable to consult your veterinary surgeon, as the breeding season is relatively short.

Your vet will try and obtain a history from you, to help decide where the problem may lie, and whether it is related to male infertility, female infertility, a combination of the two, or to a management problem such as undernutrition.

Keep breeding records! Even if you have only a few goats, jot down in your diary the dates that oestrous occurred, and if the goat was served (and by which buck). Its previous kidding date, and any known kidding problems the previous year – such as assisted kidding, retained afterbirth, metritis – may also be useful.

Female Infertility

Infertility in the female goat is complex due mainly to the interaction between the ovaries and the pituitary gland at the base of the brain, which in turn is influenced by a range of environmental and other factors. The photograph shows a non-pregnant uterus and ovaries. Female infertility can be investigated according to the following symptomatic categories:

- Problems at service:
  - Doe not on heat.
  - Doe apprehensive/nervous (this can occur with a maiden doe and a large buck, for example).
  - Vaginal abnormality, for example a constriction.
- Absence of heat (anoestrous):
  - Doe pregnant.
  - Doe not actually exhibiting outward signs of oestrous, or owner not detecting heat.
  - Is it the goat breeding season?
  - Malnutrition.
  - False pregnancy ('cloudburst').
  - Ovarian disease or congenital abnormality, for example 'freemartin' (intersex).
- Irregular cycles (long):
  - Embryonic death – a doe conceives (and thus stops cycling), but the embryo dies, and the doe returns to normal cycling activity.
  - Silent heat – the doe is cycling normally, but not exhibiting outward signs.

- True anoestrous – a total cessation of normal cycling due to an abnormality of the ovary (often referred to as a 'persistent corpus luteum' requiring hormone therapy from your vet to control).
- Irregular cycles (short):
  - 'Follicular cysts' – these can develop in the ovary, causing apparent, though irregular periods of oestrous, but no egg is available to fertilize. These require hormone therapy from your vet.
  - Irregular short cycles can develop quite naturally at the beginning and end of the breeding season, and in the first heats shown by maiden females – and this is quite normal.
- Regular cycles (not conceiving):
  - Infertile male, or problems with insemination procedure or viability of semen used.
  - Doe presented for service when not in oestrous.

Male Infertility

- Overuse: A comfortable ratio of buck/does is 1:70 for a mature buck, and 2:70 if they are young bucks. The ratio may need to be lower if the does have been synchronized, when a high proportion may be on heat on any one day, thus resulting in a high service workload for the buck.
- Failure to serve:
  - Low libido.
  - Foot problems, for example footrot, laminitis (*see* Chapter 11).
  - Arthritis.
  - Back pain.
  - Abnormalities of the penis/prepuce.
  - Generalized illness.
- Poor semen quality – a successful service appears to have occurred, but the does fail to conceive:
  - Overuse.
  - Age – old or immature bucks may produce semen of a lower quality.
  - Semen quality low at the start of the breeding season.

- Poor nutrition – deficit in energy/protein intake; good fertility relies on the buck being on a rising plane of nutrition during the early breeding season, which should be maintained. Beware parasites that could cause gut malabsorption and loss of weight.
- Other systemic disease, particularly one causing a pyrexia (raised temperature).
- Abnormalities of the testicles.
- Heavy fibre covering of the scrotum in fibre goats has been associated with a raised scrotal temperature and reduced sperm density. Even after shearing it may be 6–8 weeks before normal fertility will return.

If you suspect that your buck is suffering from any of the above conditions, your veterinary surgeon should be consulted, when a full clinical examination can be carried out to find the cause – this is important if the buck is to be used again before the season ends.

Bucks can occasionally develop a lactating udder, and this may hamper normal sexual activity!

Intersex/Pseudohermaphrodite

Occasionally goat kids are born that survive to maturity as an intersex – that is, they show both male and female characteristics. This can be related to polledness (born without horns), since the dominant gene for polledness is associated with a recessive gene for intersex; it follows therefore that an intersex is normally polled, with two polled parents. Affected goats are genetically female, with a normal female chromosome complement (60XX), but show much variation in the sexual organs. The worst affected kids are obvious at birth, with a vulva but enlarged clitoris (often referred to as a penile clitoris). The gonads may be testes or 'ovotestes', and may be abdominal or scrotal. Less severely affected kids may show few abnormalities and grow to maturity, when they may not be noticed until they fail to conceive.

PREGNANCY

The average gestation length (pregnancy) in the goat lasts for 150 days. Pregnancy can be confirmed by a number of procedures:

1. Absence of oestrous – although oestrous will normally cease once a doe is pregnant, this is not a particularly reliable method of pregnancy diagnosis. Many does, particularly in small groups, show few outward signs of oestrous anyway, and the seasonal breeding pattern may lead to irregular cycles, particularly at either end of the season.
2. Udder development – again, not a reliable method, particularly as udder development can develop in non-pregnant goats (so-called 'maiden milkers').
3. Abdominal distension – again, not reliable. Abdominal distension can develop for a number of reasons, including false pregnancy (cloudburst) – *see* later in this chapter.

4. Oestrone sulphate assay – this milk or blood test is quite widely used, and is fairly accurate after fifty days of gestation.
5. Progesterone assay remains high throughout pregnancy, being secreted by the corpus luteum. However, this same hormone can also be detected in normal cycling goats, and in false pregnancy. A sample taken twenty-four days after a true mating will give nearly 100 per cent accuracy in determining non-pregnancy, but only 80 to 90 per cent accuracy in determining pregnancy because of factors such as embryonic death. A low progesterone level almost always indicates a non-pregnancy.
6. Ultrasound scanning – in skilled hands this is a very accurate means of assessing pregnancy, giving the added benefit that the number of kids carried can also be assessed. The technique can be used as early as twenty-eight days post service, but due to the risk of embryonic loss occurring, it may be better to wait until 50–100

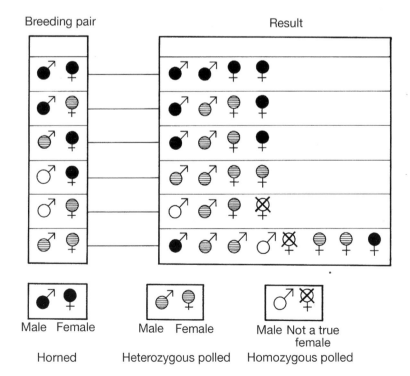

Intersex frequency.

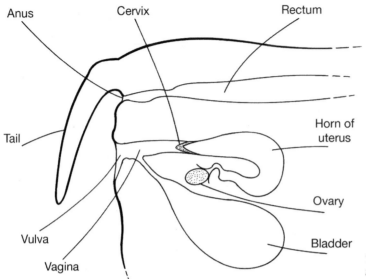

The reproductive organs of a non-pregnant doe.

days before carrying out the procedure.

7. X-rays will show foetal skeleton development from 80 to 90 days of gestation.
8. Foetal ballotment of the right flank may detect a foetus, or evidence of foetal movement in the last thirty days of gestation.

False Pregnancy ('Cloudburst' or 'Hydrometra')

This is a very specific and well defined problem in the goat, and has become quite widely recognized in the past few years, with an apparent increase in incidence as a result of the manip-

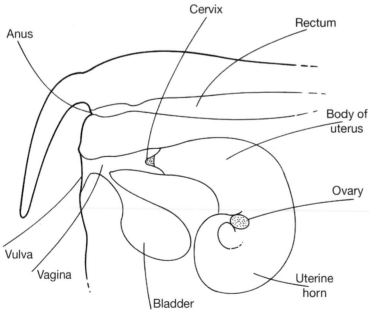

The reproductive organs of a pregnant doe.

ulation of day-length and artificial breeding in commercial herds. The condition can potentially develop in any doe of breeding age.

Clear fluid builds up in the uterus in the absence of conception and pregnancy, but in the presence of a persistent corpus luteum, which continues to secrete progesterone, thus giving all the hormonal signals of pregnancy. It may also follow a successful conception, but one that has resulted in early embryonic death.

The condition tends to occur more commonly in older females, in which the incidence in a naturally breeding doe (during a normal breeding season) will be somewhere between 2 and 3 per cent, and this may rise to 8 per cent in an out-of-season breeding programme.

The doe acts as if pregnant, with enlargement of the abdomen and some udder development if not already milking. One complicating factor is a sharp drop in milk yield if the condition is not recognized and treated.

The fluid produced gradually increases in quantity (up to 7ltr/12 pints), and the abdomen becomes grossly enlarged. If the false pregnancy follows embryonic death, then it may persist for the full gestational length or even longer before luteolysis (destruction of the corpus luteal tissue) occurs and progesterone secretion ceases, and the foetal fluids are released (hence the term 'cloudburst'). Many affected does will milk adequately following this event, as if their lactation is following a normal pregnancy.

If the false pregnancy was not linked to embryonic death, then the duration of the condition can be very variable. Although affected does can conceive naturally in subsequent seasons, there is a tendency for the condition to recur.

The fluid produced is usually clear and sticky, but it may contain placenta and remnants of foetal tissue if it follows a successful pregnancy.

The condition must be differentiated from a normal pregnancy by the means outlined previously. Your veterinary surgeon can terminate the condition if required by the use of a prostaglandin (to initiate luteolysis) and oxytocin (to aid in the expulsion of the fluid). Following treatment, approximately 50 per cent of affected does will become pregnant by 100 days if served.

The condition also needs to be differentiated from the abnormal pregnancy referred to as 'hydrops amnios/hydrops allantois', in which a similar fluid can build up in the developing placental tissue in a goat that is also pregnant; up to 10ltr (18 pints) of fluid may be produced (*see* later).

ABORTION

Up to 2 per cent of all pregnancies in the goat may result in a visible abortion (you actually see dead premature kids expelled before they are due), but this does not include those earlier pregnancy losses in which the embryo dies and is reabsorbed. It is likely that many of this 2 per cent of kids can be described as 'nature's losses' due to abnormality of the developing foetus or its placenta.

Such a figure is meaningless, however, to an owner with only a few goats, and on small units any abortion should be investigated as a matter of urgency. As goat herd numbers increase, then the occasional abortion may be of less concern, but more than two abortions in 100 pregnancies, particularly over a short time-scale, may indicate a developing problem.

Abortion caused by an infectious agent can cause a rapid increase in abortions (often referred to as an 'abortion storm'), and not surprisingly this type of abortion occurs mainly (but not exclusively) on larger units. There are a number of reasons for this:

- Goat units that expand by purchasing breeding does may inadvertently 'buy in' infection in a carrier animal. Your herd biosecurity is discussed in Chapter 3.
- Housing groups of goats together will increase the likelihood of infection spreading.
- The careless management of an outbreak

by herd owners may result in infection being spread between groups.

- On a unit with low profitability, 'cost cutting' may result in the herd manager failing to investigate an abortion outbreak properly, so that suitable control measures (although available) are not adopted.

Human Health Importance

Almost all the infectious agents that can be associated with abortion in goats (and co-incidentally in sheep) have the potential to cause illness in humans. Strict personal hygiene is of paramount importance, so you should always ensure that:

- Disposable gloves are worn to handle the products of abortion; if gloves are not worn, then hands must be washed thoroughly.
- You do not eat or drink or smoke if you are handling aborted material (or normal kidding does, since not all goats will abort, and infection can be carried to full term with kids and placenta contaminated).
- You wear overalls or old clothes when working in a kidding area, and that you don't wear these in the home environment.

Pregnant women should avoid contact with aborting goats and sheep, since some agents – particularly one referred to as causing enzootic abortion – can be dangerous to the developing baby and to the health of the mother.

Remember to exclude visitors to your goats who may be pregnant. Premises where goats are kept in areas where the general public visit may wish to display a sign warning of the potential dangers.

PREGNANT WOMEN ARE ADVISED TO AVOID CONTACT WITH SHEEP AND GOATS.

The Infectious Causes of Abortion

Abortion is most commonly caused by:

- enzootic abortion, often referred to as EAE, or 'enzootic abortion in ewes' – the main species it infects;
- toxoplasmosis;
- listeriosis;
- campylobacteriosis;
- Q fever.

Less common causes include salmonellosis, and border disease.

Non-Infectious Causes of Abortion

It is always difficult to 'prove' a non-infectious cause, but the following have been blamed for causing abortion:

- Being chased by dogs or children.
- Rough handling/getting caught in, for example, gates or hedges.
- Being transported, for example, to and from a show.
- Malnutrition – although malnourished goats in underdeveloped countries will readily carry kids to full term.
- Vitamin/mineral deficiencies – again, there is little proof to support this.
- Other illness, particularly any infection that causes the doe to be pyrexic (run a high temperature).
- Following medication – veterinary surgeons may administer a corticosteroid to a sick goat, and such a product can cause abortion as a side effect. Your vet will almost certainly ask you if your goat is pregnant, but make sure you always tell your vet no matter what medication they are administering to be safe. Equally, always ensure that you read the label of any medicinal product that you administer; many will carry the phrase: 'It is inadvisable to administer this product to a pregnant animal', or something similar.

Abortion Action Plan

Be prepared! If you discover that a doe has

aborted, assume the worst, that it may indicate a developing infectious storm, and do everything you can to stop it spreading. It will take a few days for a laboratory to confirm a cause, so it is better to be safe than sorry!

1. Isolate the aborting doe or does from all other goats.
2. Record their eartags/names in your diary, and in larger herds, consider marking them with a permanent marker so they can be easily identified.
3. Collect up all the products of abortion that you can find, including kids and placenta, and place them in a suitable leak-proof container such as a plastic bag. Fork out any bedding that is visibly contaminated, and burn or bury it. Make sure that any aborted material not required for laboratory examination is disposed of sensibly; it can be placed in, for example, a plastic bin to prevent it being dragged off by dogs, cats or other wild animals.

A dedicated 'wheely bin' makes an ideal container to prevent aborted material being dragged off by dogs, cats or other wild animals.

4. Seek veterinary advice to help you find a cause. Your vet will probably examine the material and select suitable samples to send to a laboratory, and a blood sample from the dam may also be useful.
5. When samples have been selected and despatched to the laboratory, ensure that the remaining material is disposed of sensibly.
6. Continue to be vigilant: remove further aborting does into isolation, and gather the aborted material for disposal.
7. Human health – remember to keep pregnant women away from the kidding/aborting area and products of abortion. It also makes sense to keep other potentially vulnerable individuals away, including the very young, the elderly, or anyone on medication that may have damaged their immune system.
8. Ensure that everyone in contact with the abortion incident washes their hands properly after handling goats or potentially infected material, including fixtures and fittings.
9. Do not allow anyone to eat, drink or smoke in the kidding area.
10. It is useful to put a disinfectant footbath outside the goat pens, the main goat building(s) and isolation pens, and ensure that everyone dips their boots.
11. Do not be tempted to 'foster' kids intended for future breeding replacement on a doe that has aborted. Infection with the agent that causes EAE can be picked up by young kids, and it can remain dormant in them until their first breeding season.
12. Equally, if any kids survive (when litter mates are born dead), then don't be tempted to keep them as replacements (particularly if EAE is confirmed), for the same reason as the preceding point.
13. After aborting, a doe will continue to excrete any infectious agent present for a few days and even weeks after aborting. Once any discharge has cleared up, they are probably 'safe', but continue to keep them separate if you can.

Samples to Send to the Laboratory

Unfortunately, as many as 50 per cent of all samples sent for laboratory examination may fail to result in a diagnosis. This may, of course, simply reflect the fact that the abortion is not caused by an infectious agent, but it may also reflect a poor or inadequate sample for the laboratory to test. Ideally send:

- The aborted kids – if you have more than one, send them all, and let the laboratory select the most suitable ones to test.
- Placenta – although in goats this is not always easy, as many goats consider it a 'delicacy', and readily eat it. If anything, this is the most valuable part of a laboratory submission, and will in fact yield more information than an examination of the kid. Spending time trying to find it may well be rewarded by a positive laboratory diagnosis (although you may well not want to hear this result!).
- Your vet may decide to take a blood sample from the doe, and may return two weeks later for a second sample for the laboratory to check for rising antibody levels to potential causes of abortion.

If you have a large number of goats, and the abortions continue or your control measures appear not to be effective, it is advisable to monitor the situation by submitting the occasional foetus or placenta. Remember, there are many causes of abortion, and it may be that more than one agent is active in the herd.

Useful Information for your Vet

- Number of does in the herd.
- Expected kidding dates.
- Stage of pregnancy at which abortion(s) have occurred.
- Number of abortions to date.
- Are the does showing any signs of illness – in listeriosis, for example, other signs may be apparent, such as nervous disease.
- Age(s) of does aborting.
- Any vaccines being used.

- Current feeding – listeriosis, for example, may be related to spoiled or poor silage.
- Any recent stress, such as handling/transport.
- Any medication given recently.

COMMON CAUSES OF ABORTION

Laboratory tests may confirm any of the infectious causes already described. This section will give the reader some background information about each condition, but it is important that you consult your vet on current and future control measures.

Note that although many of the infectious causes of abortion in goats can also cause abortion in sheep (and vice versa), there are some subtle differences in presentation and control that are important. So even if you have both sheep and goats on your farm or smallholding, your approach may well be different.

Chlamydial Abortion (Enzootic Abortion)

Cause: *Chlamydophila psittaci* (formerly *Chlamydia psittaci*), a modified/adapted bacterium.

Human health significance: The causative organism poses a risk to human health, and particularly to pregnant women. The aborted kids, placenta, foetal fluid and any continuing vaginal discharge are all highly infectious. The milk from an infected doe may also be infected, particularly if the udder and teats become contaminated.

The risk is greatest in the goat pens, but care should also be taken when washing clothing or overalls potentially contaminated with the products of either abortion or of normally kidding does – again, pregnant women should avoid handling this material where possible.

Live but weak/sickly kids born during an outbreak of abortion may be infected, and pose a threat if they are brought into the household area: this is best avoided altogether. Avoid using hairdryers/fan heaters to warm kids up:

this can be a particularly hazardous procedure, especially in an enclosed environment, since a steam or aerosol can be created in which droplets could carry the infectious agent that can then be inhaled by anyone standing nearby.

The introduction of infection: The most likely way for infection to get into a clean herd is by the purchase of an infected carrier female. It is possible that contamination on clothing/boots could spread infection if the wearer had been helping out with kidding on a friend or neighbours farm. As the organism also causes infection in sheep, then spread between sheep and goats is a possibility either by direct contact or indirectly via clothing.

The author investigated one outbreak of Chlamydial abortion in goats on a farm that had recently suffered an outbreak of EAE in the sheep flock. It was thought that foxes had dragged infected placenta/aborted lambs into the goat building.

Clinical signs: Abortion can occur at almost any stage of pregnancy (unlike sheep, in which abortion tends to occur in the last few weeks of pregnancy). The incubation period is very short (as little as 2–3 weeks) and infection can spread rapidly within one kidding season. This is very different again to sheep in which infection tends to be picked up one year, with abortion occurring the following year.

The placenta is normally thickened and congested (and a deep red colour), often with visible pus over its surface. The kids may look totally normal. The doe usually shows no clinical disease.

What can I do?

1. In sheep, a long-acting injection of an antibiotic (usually oxytetracycline) has been shown to be effective if given to the remainder of the group at risk. This may be worth trying, but be guided by your vet.
2. Segregate any goats that abort, and keep them separate until all discharge has dried up.
3. Consider vaccination to protect the remainder of the goats. As already stated,

the pattern of spread is different to sheep, and it may be useful to protect does that are to kid later in the breeding season, and also in subsequent years. There are a number of commercial vaccines available; discuss the best option with your vet.

Toxoplasmosis

Cause: This is a protozoal (single-celled) parasite (*Toxoplasma gondii*) that has a complex life cycle involving at least two different hosts. The diagram shows the Toxoplasma life cycle, with the cat as the main host, and goats (and more commonly sheep) as the secondary host, resulting in abortion or stillbirth or the birth of weak, congenitally infected kids. The secondary host can also include man.

Human health significance: Although Toxoplasma can infect humans, it is not generally considered to be the goat (a secondary host) that acts as the source of human infection (also a secondary host), but the primary host (the cat). The faeces of infected cats are rich in the infective Toxoplasma oocyst, and the risk comes from, for example, cleaning out cat litter trays, with pregnant women again advised to observe caution.

A Toxoplasma developmental stage (a tachyzoite) may be passed out in the milk of infected does, and although posing only a slight risk (as the human stomach acid will kill this particular stage), it may be advisable to prevent children and pregnant women from drinking raw milk.

Introduction of infection: Cat faeces contaminating goat feed is an obvious route of infection; and hay and straw can also become contaminated if cats use them regularly to sleep on, or to give birth to, and rear their kittens on them.

Clinical signs: Toxoplasma infection picked up at any time other than during pregnancy will produce no obvious clinical disease. During pregnancy, however, the placenta and developing kids are attractive targets for the developing and multiplying toxoplasma parasites, causing cell damage and death. The results of this infection depend on the stage of pregnancy at the time when oocysts were eaten, and on the degree and extent of the ensuing damage caused to the placenta and foetal kids.

1. Infection during early pregnancy may result in early embryonic death so that does return to service without visible evidence of having aborted.
2. Infection during mid-pregnancy may result in foetal death and abortion of mummified or decomposing kids 2–3 weeks before the expected kidding date.
3. Infection later in pregnancy may result in the birth of stillborn or weakly kids at term.
4. It is possible for live kids to be born alongside mummified litter mates!

Although sheep will readily develop strong protective immunity after infection, this does not seem to be the case in goats. Susceptibility to new environmental challenge can be a problem, as can re-activation of an existing

dormant infection if the goat is stressed.

The appearance of the foetus depends on the stage of pregnancy; thus earlier abortions are often mummified, with a dry and shrunken appearance, whilst later abortions may show no obvious foetal abnormality. The placenta will occasionally show characteristic white 'pin-head'-size lesions on the cotyledons.

What can I do?

1. As cats are the main source of infection, it is important to keep them away from goat feed stores (including hay and straw) and from goat buildings. As young cats and female breeding cats are the greatest source of infection, then having farm cats neutered is a useful approach, as this will cut down the number of kittens. There is no need to get rid of the cats: some sheep farmers often attempt to do this, but other cats from neighbouring farms will usually move in. The secret is to recognize the role that cats play, and to manage them effectively.

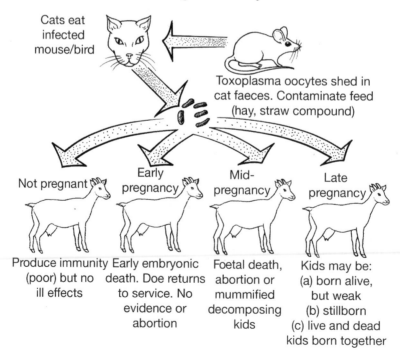

The Toxoplasma life cycle and likely outcome, depending on the stage of pregnancy.

The typical appearance of a mummified foetus (this is a calf).

Cats are a recognized source of Toxoplasma infection; their faeces can contain high levels of Toxoplasma oocysts.

2. There is no recognized treatment available.
3. Remove the products of abortion and send for laboratory diagnosis.
4. As immunity is poor, one approach is to cull infected goats to prevent potential abortions in subsequent pregnancies.
5. A vaccine is commercially available for use in sheep in many countries worldwide; although it is not licensed for use in goats in the UK, it may well have a product authorization in other countries worldwide, although any immunity imparted may well be disappointing.

Listeriosis

Cause: *Listeria monocytogenes*, an organism that can cause a wide variety of clinical signs in goats (described in other chapters of this book).

Human health significance: Human infection with this organism can occasionally occur, but animal-related incidents are normally linked to the consumption of milk products such as cheese. It is important to put into perspective the fact that *Listeria* spp. are fairly ubiquitous in the environment, with high levels potentially developing in silage or rotting vegetation. However, as the cause of abortion will not be known for sure until laboratory tests are complete, caution is urged, particularly if there is a vaginal discharge (*see* below); furthermore the milk of does that

abort may carry high levels of Listerial infection and should not be consumed.

Origin of infection: On larger units, silage (and in particular grass silage) is a common source of infection. Any spoiled feed/decaying vegetation/stagnant water is a potential source.

Clinical signs: *Listeria monocytogenes* can cause a wide spectrum of different clinical presentations, including encephalitis, septicaemia, diarrhoea, death – and abortion can develop in an incident where all, some, or none of these other clinical signs are apparent. Abortion tends to occur in the latter stages of pregnancy, and the doe is commonly 'off colour', with retained afterbirth, metritis and a vaginal discharge all possible complications.

What can I do? Discuss this with your vet, but if the doe is ill, then antibiotic medication may be needed. If the source of infection can be removed from the diet, then fresh cases may be prevented. Any doe that aborts is likely to be immune, and can be safely retained.

Campylobacter (formerly Vibriosis)

Cause: *Campylobacter fetus fetus*, and possibly *Campylobacter jejuni*.

Human health significance: Minimal, although some strains of Campylobacter can cause stomach upsets and diarrhoea – again, caution is urged, particularly before a diagnosis is made.

Origin of infection: Infection is normally introduced into a group of goats by a carrier animal (either goat or sheep, the organisms involved are common to both). This carrier animal may be perfectly healthy, shedding infection in faeces, as the organism is a common gut inhabitant. The mixing of goats in the second half of pregnancy is a particular risk factor, since healthy carrier animals may be introduced to susceptible animals that are pregnant, and abortions may then begin. Mixing infected and non-infected goats together at any other time will result in susceptible animals developing immunity, and no other clinical disease will normally be seen.

Clinical signs: Abortions will tend to occur in the last few weeks of pregnancy, although weak, sickly kids may also be born. The doe may be 'off colour', or appear totally fit and healthy. A vaginal discharge is a common sequel to abortion, and the placenta is often covered in a brownish discharge.

What can I do? Separate any does that abort from other pregnant goats. If Campylobacter infection is confirmed, it may be beneficial to mix them with other non-pregnant does or with those that have aborted. This approach spreads the infection and hence immunity through the herd. This must not be attempted, however, unless there is firm evidence to support a diagnosis (and remember that mixed infections can also occur) – the approach could be disastrous for some other infectious agents, such as Chlamydia.

Your veterinary surgeon may decide to attempt control with antibiotics if the problem begins to escalate.

Q Fever

Cause: *Coxiella burneti.*

Human health significance: Refer to advice given under Chlamydial abortion above. All products of abortion are potentially hazardous, and Q Fever is a recognized illness in humans that can result from exposure.

One well documented incident of Q Fever in the UK occurred in a family with a flock of sheep in which Q Fever had been confirmed. Although no illness was reported in those working in the lambing areas, Q Fever was confirmed in other family members living in the main farmhouse. It transpired that weak and sickly lambs were being brought into the house, and warmed with a hairdryer. The hairdryer created a steam, and many of the water droplets within the steam contained the Coxiella organisms from the skin surface of the lambs. These droplets (and hence infection) were then inhaled into the family members' lungs in the closed environment of the house, and Q Fever then developed.

Do not use a hairdryer to warm and dry a sickly lamb!

Origin of infection: Infection is invariably introduced into a herd by the purchase of infected healthy carrier does. It is a recognized phenomenon with this agent that healthy kids can be born to an apparently healthy doe, but that the afterbirth, kids and uterine fluids can be heavily contaminated with the Coxiella organism. Spread at kidding time can be rapid.

Clinical signs: Q Fever can cause abortion in the last few weeks of pregnancy. The afterbirth often has a covering of a thick brown discharge.

What can I do? Your veterinary surgeon may decide to use antibiotics if the problem is escalating.

KIDDING: RECOGNIZING AND AVOIDING PROBLEMS

Kidding time can be hard work on larger units, but intensely rewarding for any goat owner. It may also be a worrying time if you are ill prepared, and preparation is the key to success.

Earlier in this chapter we discussed the seasonal breeding pattern in the goat, and how this can be manipulated by artificial lighting. We emphasized the value of pregnancy diagnosis, and an accurate service and hence kidding date. These management procedures and the simple use of a diary, a calendar or a computer means that no one should be caught out!

One further benefit of scanning is that the number of kids carried can be estimated, thus enabling larger units to batch up pregnant does in the latter part of pregnancy, with those carrying multiple kids receiving special dietary care.

Forward Planning

- Mark the expected kidding dates on your calendar well in advance!
- Decide where your does will kid. Most will kid during the spring, and will probably be housed. If they are to kid in their own pen, make certain that the pen is clean, dry, free from draughts and well bedded. With larger groups of goats, however, it is probably better to move them to a separate pen or kidding area. Ensure that this area has been well cleaned and disinfected using an approved disinfectant at the correct concentration; an alternative is 1lb (0.45kg) of washing soda to 10gal (45ltr) of water.
- The best choice of bedding is undoubtedly straw, although wood shavings/sawdust or shredded paper can also be used if straw is not available. These alternatives do have important disadvantages however, as they are very absorbent and can easily become wet, sticking to the goat's udder and teats, and becoming impacted between the claws. This type of material should also be removed between kiddings, whereas straw, providing it is not too wet or contaminated, can simply be topped up with fresh straw.
- *No matter what approach is adopted, the kidding pen must be kept scrupulously clean at all times if newborn kids are to be protected from disease.*
- Ensure that you have a permanent or a secure temporary supply of electricity in the kidding area. Good lighting is essen-

tial, and a supply of heat such as a heat lamp may also be needed. All wiring must be kept well away from the goats, who may readily nibble and chew through it given the chance – and cases of electrocution have been recorded. If heat lamps or heaters are to be used, ensure that these are safely suspended out of reach, thus ensuring there is no chance of fire.

- Clean drinking water must be available at all times, but be careful to raise buckets or containers off the floor and secure them just in case a weak newborn kid falls in, or knocks it over and drowns.
- Provide good quality, palatable feed to does around kidding time. If the pen is temporary, make sure any containers are secured so they can't be knocked over and the food soiled. Avoid feeding hay or silage off the floor, as kids will simply use it as bedding. Equally, avoid using feeding nets within kids' reach, as they may hang themselves. Use a metal rack or similar container firmly attached to the wall.

Doe Problems Around Kidding Time

Towards the end of pregnancy, and particularly in the few days either side of kidding, the doe is susceptible to a group of conditions referred to as 'peri-parturient' (literally 'around kidding').

Pregnancy Toxaemia

Cause: The two main predisposing factors are over-fatness, and carrying multiple foetuses, and the condition tends to occur in the last 4–6 weeks of pregnancy.

Over-fatness should be avoided during pregnancy wherever possible. Problems may begin before the doe is pregnant if goatlings are overfed, but it is overfeeding from mid-pregnancy onwards that leads to excessive fat being laid down internally, as the so-called intra-abdominal fat. As the foetuses increase in size, the available rumen capacity is progressively reduced, further exacerbated by these intra-abdominal fat deposits. The result is that insufficient feed can be consumed to meet all the demands of pregnancy, and the doe then begins to mobilize this excessive fat, which becomes deposited in the liver and kidneys, resulting in liver and kidney damage and progressive failure.

Clinical signs: Affected does progressively lose their appetite (thus adding to the problem). They then become lethargic and unwilling to move, and if untreated this is followed by recumbency and ultimately death. Some affected does may show nervous signs including blindness, head pressing (standing in a corner with their head pressed against the wall), and some may adopt a 'stargazing attitude'.

Treatment: Although the clinical signs are fairly typical, your veterinary surgeon may take a blood sample to check for beta-hydroxybutyrate (BHB) levels, produced as fat is mobilized excessively. Mild cases may respond to tender loving care, and tempting with favourite foods and titbits such as warm tea with sugar. It is said that plants such as rose-bay willow-herb, ivy, dandelions, and branches from trees and hedgerows are all potentially beneficial. Your vet will probably administer glucose into the vein, although there are a number of proprietary products for sheep, sold as 'twin lamb drench', that may be helpful in milder cases. If the condition fails to respond, your vet may take the next step of inducing kidding or even carrying out

a Caesarean to remove the kids. If not recognized early, the prognosis for both doe and kids can be very poor, however. It is usually advisable to treat the doe for hypocalcaemia (low blood calcium), a common complicating factor.

Fatty Liver Syndrome (Hepatic Lipidosis)
It is normal for pregnant does to deposit some fat in the liver around kidding, as it is mobilized from body fat stores in preparation for kidding and impending lactation. However, overfeeding in mid- and late pregnancy, even by as little as 10 per cent, may result in liver cells storing fat beyond their capacity to mobilize and release it, leading to impaired hepatic function. The moral is to ensure that your goat is not too fat in the critical second half of pregnancy.

Hypocalcaemia (Milk Fever)
Cause: A fall in the level of circulating calcium and/or phosphorus in the blood, which can occur at any time in the periparturient period, but particularly in late pregnancy as the udder fills with milk ready for a new lactation. This condition is particularly common in dairy cows, and is an occasional problem in goats. A sub-clinical form is probably the most common presentation, and may accompany a range of other conditions occurring at this time, such as mastitis and metritis.

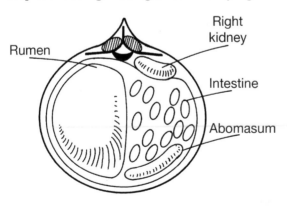

Cross-sectional diagram of the abdomen of a non-pregnant goat.

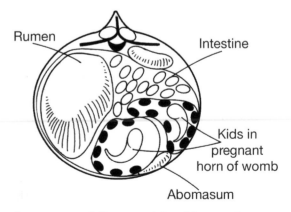

Cross-sectional diagram of the abdomen of a pregnant goat carrying two near-term foetuses. Note the reduced rumen volume.

Clinical signs: In true clinical disease, signs may include tremors, ear twitching, hyperexcitability and unsteadiness on the feet (ataxia). This may progress to fits, recumbency, coma and death if unrecognized and untreated.

Treatment: The injection of calcium (usually as calcium borogluconate) can bring about a rapid improvement. Although this can be administered under the skin over the ribs, a better response occurs if it is given intravenously, and this is best carried out by your vet.

It is recommended that all recumbent and comatose goats be given calcium.

Prolapse of the Cervix

Cause: No single factor is responsible for prolapse of the cervix, but it is a relatively common problem in pregnant does in late pregnancy. It is generally associated with an increase in the pressure exerted on pelvic content, which initiates straining by the doe thus forcing the vagina/cervix through the vulva. Specific factors identified have included:

- multiple kids;
- bulky feed and a rumen of diminishing size as the uterus enlarges;
- in housed goats, lack of exercise in late pregnancy may also be important;
- the conformation of the dam, including the shape of the pelvis;
- excessive coughing, as in an outbreak of pneumonia, may also increase the incidence of prolapses.

Clinical signs: A pink 'soft swelling' appears beneath the tail at the entrance to the vagina. It may vary from a small swelling only visible when the doe lies down (disappearing when she stands), to a large, football-sized swelling that may include the urinary bladder and intestine forced through the pelvis and contained within the prolapse. The doe will strain excessively, and may well be very vocal due to the pain and discomfort involved. The surface can quickly dry out, and become badly damaged if knocked or bumped; it may even rupture,

with intestines protruding through the torn surface.

Treatment: Once recognized, treatment needs to be rapid and gentle, but above all skilful. Small prolapses can be effectively controlled by the use of proprietary ewe prolapse harnesses that can be adapted to goats. More severe cases need to be gently eased back through the vulval lips, and this may well be a job for your veterinary surgeon who will normally administer an epidural anaesthetic to prevent the doe straining. The prolapse needs to be cleaned and lubricated before being replaced, and the most effective way of ensuring that the prolapse does not recur (and the one favoured by your vet) will be to apply an encircling suture around the vulval lips.

Unfortunately, dead kids can be a fairly common sequel, and the administration of antibiotics by your vet will often prevent infection becoming a problem if this does occur. There may also be a tendency for prolapses to occur in successive breeding seasons, so in larger commercial herds, culling may be a favoured option.

Intestinal Prolapse

As previously described, this condition can occur as a complication of a torn or damaged cervical or vaginal prolapse, but it can also occur spontaneously. Affected goats are usually found dead with a mass of intestine protruding from their back end, as a result of a severe shock. If found alive, then your veterinary surgeon needs to be contacted as a matter of urgency so that the goat can be humanely destroyed; treatment is usually hopeless.

Kidding Equipment

The telephone number of your vet should be clearly located near your main telephone, and enter it into your mobile. Have a notebook and pencil handy if you are phoning out of normal hours; you may be given an emergency number to phone.

Whether you are expecting one doe or one thousand to kid, it is important to ensure that you have a few basic items in a box or bucket that you can readily locate if problems develop. Many DIY stores stock suitable plastic containers with lids to store your kidding kit in. You should include the following:

- Disposable arm-length gloves, and latex gloves.
- Old but clean towels.
- Sterilizing fluid.
- Disinfectant.
- Soap for washing your hands.
- Hand cream – particularly important if you are kidding a number of does, or if your hands are prone to developing chaps and sores.
- A torch – preferably a head torch that allows you to use both hands, with the torch pointing in whichever direction you are looking in.
- Scissors.
- Lambing ropes and a lambing snare – proprietary ones are available at agricultural merchants. Also have at least three different lambing ropes available in three different colours, one for the head rope, and one each for the two front legs.
- Obstetrical lubricant – use commercially available products in liquid form, with a large nozzle, and preferably one that remains liquid in cold weather.
- A respiratory stimulant, available from your vets. It can be placed on the back of a weak kid's tongue if it is not breathing, and can be very effective.
- Calcium borogluconate – discuss its use with your vet. As mentioned earlier in this chapter, hypocalcaemia may be a common

sub-clinical problem, and does that are weak or unable to stand after kidding may well benefit from a calcium boost.

- Lamb stomach tubes – these are available either from your vet or from agricultural merchants. Learn how to use them before attempting the procedure yourself, but colostrum administered by stomach tube in the first few hours of life in a weak kid can be extremely beneficial.
- Stored colostrum, *see* Chapter 6.
- Syringes and needles of various sizes.
- Navel treatment – it is important that the navel of the newborn kid is kept clean, and dries up quickly to prevent potentially life-threatening infections gaining entry to the body and causing infections referred to as joint ill and navel ill, as well as more serious infections. There are two main approaches: one is to use strong tincture of iodine, and the second an antibiotic spray; the former is both cheaper and more effective, although the latter may be more convenient! Tincture of iodine can be obtained either from your vet or from a local chemist. Do not confuse it with the iodine-based teat dips sold by agricultural merchants – these will be ineffective. It is best applied to the wet navel immediately after birth by placing some in, for example, an old eggcup, and simply immersing the navel right up to the junction between navel and hair. Some shepherds will place the iodine in an old washing-up bottle to use when lambing sheep out in a field. The iodine can then be squeezed over the navel, the kid turned upside down, and the procedure repeated on the other side.
- Thermometers – it is always useful to have more than one thermometer in your kit, as they are easily broken, dropped or lost! The traditional glass thermometers are still widely used and are reliable, but increasingly now digital thermometers are appearing on the market. Whichever type you use, become familiar with it, and practise reading it, particularly by torchlight! Make sure you shake down the mercury in

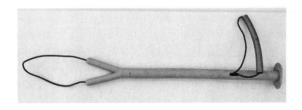

A lambing device suitable for use in kidding does.

the glass thermometer (though be careful not to do this too vigorously or you may find it flies across the shed and smashes in the corner) or re-set the digital thermometer before use. It is sensible to lubricate the thermometer before inserting it into the rectum; use either obstetrical lubricant, or vaseline (or even some spit). Insert it gently, directing it towards the rectal wall, and hold it against the wall for 1–2min. Wipe it on a tissue and read it.

- Rubber castration rings – if you are keeping your male kids, and they are not being reared as breeding animals, then rubberring castration is the best approach.
- Bottles and teats in case you need to bottle-feed orphaned or weak kids.
- A marker spray stick, invaluable for marking kids if several does kid together; simply spray the doe's number on the side of the kid as a temporary measure.

No Substitute for Experience

If you are a novice goat keeper, you must recognize your limitations before your goats begin to kid. There are many books that give advice on how to assist at a calving, lambing or kidding, including this one – but there is no substitute for experience or practical training!

Your local goat club should be able to put you in touch with other goat owners in your area who may be able to help and advise, and it may even be possible to gain some practical experience on larger goat units.

Local agricultural colleges may run lambing courses, and some veterinary practices run similar courses for their clients. Although these tend to be dedicated 'lambing' courses, many of the principles involved can be readily applied to a goat that is kidding.

Signs of a Normal Kidding

A good stockperson will recognize the often subtle signs that kidding is imminent. The doe will look anxious and apprehensive, and will stand quietly in the corner of her pen. As time passes, she will often begin pawing the bedding with her front feet, lie down, then get straight back up again. Her udder will have been getting larger over the last few days, and colostrum may begin to dribble from the teat ends.

The pelvic ligaments around the tail begin to slacken, the vulval lips will swell, and clear mucous begins to hang from the vulva, as the mucous cervical plug begins to dissolve.

First Stage Labour

As the uterus begins to contract, the foetal membranes (soon to become the afterbirth) are pushed against the cervix, the resultant pressure causing the cervix to dilate. Failure of the cervix to open is referred to as 'ringwomb' and although it appears not to be as common as it is in sheep, nevertheless kidding appears to come to a standstill as a result. As stage 1 proceeds however, and when the cervix is fully dilated, the uterus and vagina appear to merge into one, and the cervix cannot be identified.

This stage occupies between three and twelve hours, and may well not be noticed.

Second Stage Labour

When the cervix is fully dilated, the foetal membranes begin to protrude through the vulval lips and the 'water bag' is readily identified. As straining continues, this bag ruptures, and discharges clear fluid. This is quickly followed by the appearance of the kid in the birth canal, and even more intense straining will begin, until the kid is born. This stage should occupy no more than about two hours; it should be a progressive procedure, and any delays should be investigated.

As soon as the kid is born, it should be placed close to its mother's nose, and she will normally begin to lick and clean it; this is also part of the maternal bonding procedure between mother and offspring. If you are supervising the birth, make sure that any remnants of the afterbirth covering the mouth or nostrils are removed; the doe will normally do this herself, but if the kid or mother is weak, or if two or three kids arrive together, then you may have to help! Gently rubbing the newborn kid with a handful of straw or a clean towel will also be helpful in stimulating it to breath properly. The kid should quickly get to its feet; help it to locate a teat by expressing some milk from each teat first – and within a few hours the youngsters will be jumping around the pen!

Young goats kidding for the first time may initially shy away from their newborn kids, but they will usually accept them after a few minutes, and begin to mother them. Another potential problem with young inexperienced does is that they focus on the first kid born, only to neglect the second and subsequent kids. Some freshly kidded does seem to get confused if they are kept with other freshly kidded does, not being sure which kids are theirs, for example. If this appears to be a problem, then pen such a goat on her own for twenty-four hours with her kids, to allow the bonding to develop. For further advice on caring for newborn kids and preventing problems developing, refer to Chapter 6.

Make sure there are no more kids left behind; this can be done by 'balloting' – pushing into the abdomen gently with your fist, just in front of the udder.

Third Stage Labour

This involves the final expulsion of the foetal membranes or afterbirth (also referred to as the 'cleansing'). This should occur within 2–3 hours, and if it does not appear, or is hanging, it may be a sign of hypocalcaemia, and some calcium borogluconate under the skin may help. If it still doesn't appear, don't be tempted to pull it, but contact your veterinary surgeon.

CAUSES OF ABNORMAL KIDDING

An abnormal birth process in any species is referred to as dystocia (literally a difficult birth). As a result, we refer either to a foetal dystocia, when the problem is related to the kid; or a maternal dystocia if it is a doe problem. Foetal dystocia can be caused by:

- A large, oversized single kid.
- A deformed kid.
- Dead, dry kids, often with gas developing under the skin.
- Multiple births, when more than one kid is presented at the same time.
- A malpresentation, when the kid is not presented correctly; this may be a twisted head or a breech birth, for example.

Maternal dystocia can be caused by:

- A small pelvis (immature doe), or a deformed pelvis, either congenital or following an injury.
- Over-fat does, with excessive fat in the pelvis.
- Failure of the cervix to dilate – i.e. ring-womb.
- Failure of the vulva to dilate (this can be a problem in first-time breeders).
- Uterine inertia, when the uterus simply fails to contract; this can be linked to, for example, hypocalcaemia, or to illness.
- Uterine torsion – not common in goats, but it can occur, when the uterus literally twists around its axis, effectively closing down the birth canal. This will completely stop the birth process, and is a problem that you or your vet may identify by manual examination, as the birth process will come to a halt.

If the doe is straining for more than an hour without making any progress, and particularly if nothing appears at the vulva, then something may be wrong, and she should be examined. If you feel competent to do this yourself, then prepare to carry out a vaginal examination; if you do not, then give your vet a ring for advice. If you do feel competent, then approach your examination as follows:

- Make sure your hands are clean, and that your nails are not too long – they can cause

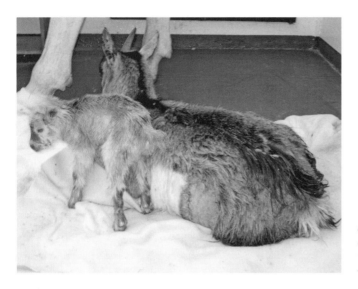

This Pygmy goat was only 9 months old at birth (following an accidental mating by its brother). The kid was born alive by Caesarean section as the pelvis was too small for normal delivery.

considerable damage to the soft vaginal tract you are examining.

- Don't forget to remove any rings, watches or other jewellery.
- Whether to use a disposable arm-length glove or not is a personal choice.
- Ensure that the vulval area of the doe is gently washed with warm soapy water if it is soiled.
- Ensure that the doe is adequately restrained: get help to hold her, preferably, or tether her.
- Lubricate the hand and lower arm or glove with a suitable lubricant from your kidding box.
- Cup your hand (fingers and thumb in a cone shape), and gently insert your hand to examine what the problem may be.
- Remember that goats can be very vocal when examined.

You may find:

A normal presentation: You need to assess whether there is sufficient room available for the kid to pass through the birth canal; it may be that more time is required. If the kid is too big, or the birth canal too small, then seek professional help – do not be tempted to pull too early in the birth process, as you can easily cause severe damage.

Ringwomb: You need to be confident that the partly opened cervix that your hand comes up against is a ringwomb, and not a cervix that is slow in opening (and simply requires more time). A competent, skilled stock-keeper can gently open a ringwomb cervix by literally wiggling their fingers around the ring of the cervix, but it can easily tear, and unskilled intervention can easily result in severe haemorrhage and death. If in doubt, seek veterinary help.

Head and one foreleg: The retained leg needs correcting before delivery in most kids, although very small kids may be born in this position.

Head only, i.e. both legs retained: If the head has been pushed through the vulva and the kid is still alive, it may be very swollen and veterinary help is required urgently, although you may be able to gently insert your hand down the side of the neck to locate at least one limb. If the head isn't too swollen, then try to gently push it back (placing a head snare or rope around it first), thus giving more room to attempt retrieval of the missing legs.

Limbs only: You must now decide whether you have front limbs or back limbs, and also if more than one kid is presented – that is, if the limbs presented belong to the same kid, or a different one. Look at the limbs of the doe: this

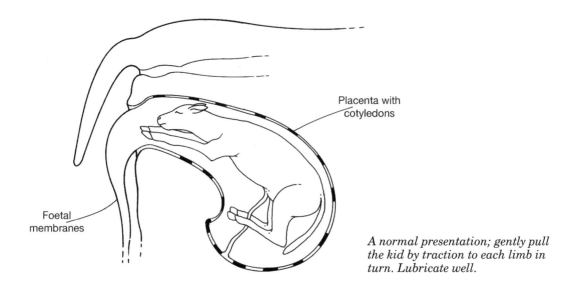

Placenta with cotyledons

Foetal membranes

A normal presentation; gently pull the kid by traction to each limb in turn. Lubricate well.

will help you decide whether you have a front or hind limb. Front limbs bend in the same direction (like our wrist and elbow), whereas the hind limb bends in the opposite direction.

- *Forelimbs only (same kid)* – locate the head, and deliver the kid.
- *Hind limbs only (same kid)* – often the birth canal has not opened up enough to deliver

it easily. It may help to turn the kid gently through 90°, followed by gentle traction. In this type of delivery, kids are more vulnerable, rib damage may occur, and they are often starved of oxygen, as the head is the last to pass through the birth canal, and the umbilical cord will have broken.

- *One hind limb only* – the second hind limb must be corrected before delivery.

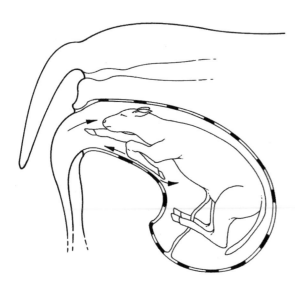

Head and one foreleg; the retained limb must be corrected unless the kid is very small.

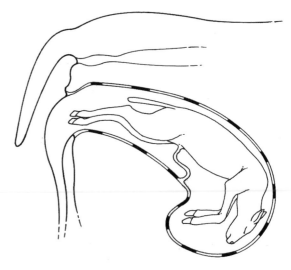

If a single pair of hind limbs can be identified (and they belong to the same kid), then apply gentle traction, twisting slightly if necessary.

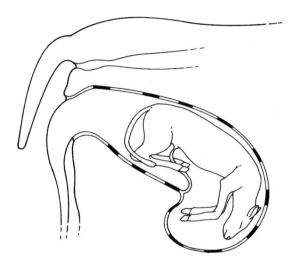

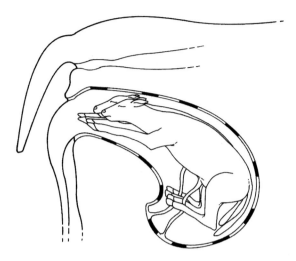

The kid must be gently pushed back, cup hands over the foot and gently pull, pushing back on the hock at the same time. Repeat with the opposing limb. Seek urgent veterinary help if in doubt.

Once you decide that you have two kids presented together, deliver them separately, by pushing back on one and applying gentle traction to the other (ensuring that both legs belong to the same kid).

- *Hocks only* – this is referred to as a breech birth. Correction is not easy without causing damage to the doe, therefore seek veterinary help if you are concerned.

Simultaneous presentation: If you decide that two or more kids are being pushed through the birth canal together, then decide which one is 'ahead' and gently repel the second kid, ensuring that you then pull only on limbs or head belonging to the same kid. Although not common, Siamese twins are occasionally encountered. If you have any concerns, then contact your own vet.

Examine with Care

Whatever you do, be very gentle: the uterus, vagina and vulva are all very delicate, and are easily damaged. Use plenty of lubrication, and do not persist in any procedure for longer than five minutes at a time, particularly if you are not making any progress. There is no disgrace in giving up and seeking professional help and advice.

VETERINARY PROBLEMS AFTER KIDDING

Once the doe has produced her kids, she has just experienced one of the most stressful insults she will receive in her life. Problems after kidding are fortunately uncommon, but watch out for the following:

Prolapse of the uterus: Fortunately this is not a common problem, but nevertheless if it does occur it is a genuine veterinary emergency, as the entire womb literally everts and protrudes from the vulva. Once you have phoned the vet, make sure other goats are kept well away, as they may inadvertently tread on the womb and badly damage it. Get a clean towel or cloth, and gently wrap it around the womb, then pour a small amount of warm water over the cloth to keep the womb warm and moist. Death from shock or internal haemorrhage is always a risk.

Rupture of the uterus: A doe that deteriorates rapidly after kidding may be haemorrhaging from a torn or damaged uterus. Blood clots may be passed out from the vulva, and the mucous membranes will be very pale.

Prolapse of the cervix/uterus – a genuine veterinary emergency.

Urgent veterinary help is needed if the doe is to be saved.

Retained afterbirth/cleansing: Goats normally shed the afterbirth quickly after birth. It can quickly rot in the uterus, and if it, or any small part of it, remains, it may make the doe very sick. Retention is not a common problem, however, so it is probably best to seek veterinary advice – one possible cause is a dead kid remaining in the womb.

Metritis: Infection of the womb can be a sequel to a normal kidding, but is more common following one in which human intervention has taken place. There will often be a foul-smelling discharge from the vulva, and the doe may run a temperature and become very sick. Consult your vet, who may place antibiotics directly into the womb in addition to antibiotic by injection.

Nerve damage: If a large kid is delivered through a small pelvis, particularly if excessive force is used, or if a doe is recumbent for any length of time with a kid partly stuck in the birth canal, then nerve damage can occur. The two nerves most commonly affected are the obturator nerve (the doe will tend to do the 'splits' when she stands and walks, particularly on a slippery surface), and the peroneal nerve (causes the hind limb to knuckle forwards at the fetlock).

Mastitis and problems of the udder: *See* Chapter 14.

CHAPTER SIX

Problems Encountered in Young Kids

After a successful pregnancy and birth, the newborn kid makes its entry into the world, and immediately a whole new range of problems need to be anticipated and prevented where possible. There is nothing more disappointing than having a well developed and perfectly formed kid born alive, only to watch it die in the first few hours or days.

FAILURE TO BREATHE

A kid must breathe to survive; it follows therefore that close supervision will recognize those kids that are not breathing, when artificial respiration may help (*see* below). Some kids are more susceptible than others – pay particular attention to:

- **Breech or backwards presentations:** These kids may inhale uterine fluid deep into their lungs, particularly if the birth is prolonged.
- **Large single kids:** Such kids may suffer damage to the ribcage during their passage through the birth canal.
- **Weak and premature kids:** This is a risk, particularly if there is a pre-existing abortion problem, such as Toxoplasmosis.
- **Multiple births:** The doe may focus on the first or second kid born, only to neglect those kids born later.
- **Young inexperienced does:** These may

not know what to do, and may walk away from the kid, simply ignoring it.

Methods of Revival

Rubbing the kid: This action simulates the rough surface of the doe's tongue and mouth, with which she will lick and nibble the kid in an attempt to revive it. This is achieved by gently rubbing your hand from head to tail over the ribcage in a gentle but fairly vigorous manner. A handful of straw or a clean towel will provide more friction.

Swinging the kid: Gently hold the kid upside down by its hind legs, and swing from side to side like a pendulum. Don't be too enthusiastic, and certainly don't swing it through a 360° arc – and don't let it slip out of your hands, as it will still be very slippery!

'Respiratory stimulants': These are obtainable from your veterinary surgeon, and come in small bottles with a number of brands available. The principle is to place a few drops (following the manufacturers' recommendations) on to the underside of the kids' tongues. This should make the kid gasp, thus drawing air deep into its lungs and hopefully stimulating the kid to continue breathing.

Stimulate a sneeze: This is usually achieved by pushing a piece of straw up the nose, gently 'tickling' the lining. As the kid sneezes, it again draws air into its lungs, stimulating its breathing.

'Mouth-to-mouth' resuscitation: But never attempt this procedure directly! One possible reason why a kid may struggle to breathe is infection, and in particular from the group of diseases that can cause abortion earlier in pregnancy, such as Chlamydia, Coxiella and Toxoplasma (*see* Chapter 5), all of which can potentially cause illness in humans. One way to avoid this risk, yet save your kid, is to use a stomach tube to get air safely into the lungs. Hold the kid as if you are going to stomach tube it (*see* later in this chapter), but pass the tube only as far as the back of the mouth (hold the tube against the side of the kid's jaw to get a rough idea how far the tube needs to be inserted, and mark the approximate distance on it with a pen or marker).

At this point, the tube will either pass into the trachea or into the oesophagus, although you in fact don't want it to enter either. Use your finger and thumb to gently pinch off the oesophagus by applying pressure behind the jaw. The oesophagus lies above the trachea, and the latter can be recognized by the cartilaginous rings that prevent it from collapsing. With your other hand, now block off the nostrils and mouth, and then gently blow through the tube. If the procedure is successful, you will see the ribcage gently rise as the chest expands. Allow the air to escape, and then repeat the procedure a few times until the kid begins to breathe on its own.

A FIT AND HEALTHY KID

Once you are sure that the kid is breathing properly, quickly check it over for any obvious abnormalities. True congenital abnormalities are rare in goat kids, but look out for:

- cleft palate;
- imperforate anus (no exit from the rectum);
- contracted tendons (limbs twisted that can't be straightened);
- a swollen neck that may indicate a possible goitre.

These are described in more detail later in this chapter.

Some goat owners will weigh the kid at birth and at weekly intervals to check its growth rate. Birthweights in kids can be very variable, ranging from over 7kg (15lb 7oz) for a single male, down to only 2–3kg in multiple births (*see* Appendix II). The larger the kid at birth, the quicker its early growth rate will be. If a number of low birthweight kids are born, then it may be worth reviewing the nutrition of the dam through pregnancy; perhaps the does themselves are in poor condition. It is recognized that under-nutrition during pregnancy can result in poor placental development, and hence poor transfer of oxygen, electrolytes and nourishment to the developing kid. One other possible reason is intrauterine infection with one of the known abortion agents; at a later stage of pregnancy, the placenta will be damaged, but if the kids survive they will be of a low birth rate.

If a kid is born up to fourteen days premature, it should have a good chance of survival, and kids up to twenty-one days premature may also survive if they are recognized and cared for.

What Should I do Next?
A fit, healthy kid should very quickly be attempting to stand and find a teat to begin suckling. Ensure that you stay with the kid to ensure that it gets its vital first feed of colostrum. If it can't stand, there may be something wrong: swayback (copper deficiency), Border disease, or spinal or limb injuries resulting from trauma during the birth process are all possibilities (*see* later in this chapter).

If you haven't already done so, check the udder to make sure that the doe has produced milk/colostrum, and that she has 'let the milk down' – the birth process itself will normally result in a good milk let-down, and milk may begin to leak from the teats. Make sure the doe has not got mastitis; if she has, the udder will feel firm and will be painful, and the milk may well contain clots (Chapter 14). Gently strip some milk from each teat to check.

Hopefully the kid will have been born into a clean and dry environment, but to further

A kid born with two heads. A rare occurrence, and the kid was destroyed on humane grounds.

decrease the risk of infection entering the kid's body via its navel, it is important to use an astringent such as iodine to dry and shrivel the navel quickly. A wet, poorly healing navel can readily become infected, leading to navel ill, joint ill and septicaemia. The author favours the use of tincture of iodine (and this is distinct from the iodophore products sold

Healthy kids should soon be on their feet suckling.

as teat dips or sprays), as it is readily available from your vet or local agricultural stockists. It can be applied by a variety of means, but the most effective is to use an old egg cup or a container of similar size, fill it with the iodine, and dip/immerse the entire navel until the hair on the abdominal wall around the top of the navel is just touching the surface – you should finish up with a 'brown colour' on this hair.

If you have a number of does kidding together, and particularly if they are all the same colour, such as the white Saanen breed, it is a good idea to mark the kids so that you can relate littermates to each other and to their dam. Coloured marker sprays from any agricultural supplier are ideal for giving this temporary mark before an eartag or tattoo is applied; the kid can also be readily identified without the stress of catching and handling.

COLOSTRUM

For a goat kid to stand a good chance of survival, it must consume an adequate amount of colostrum *within the first six hours of life*. Failure to do so can lead to malnourishment and an increased susceptibility to disease.

69

So Why is Colostrum so Important?

Colostrum is the first milk produced by the doe (and most other mammalian species), and is an essential source of antibodies in the form of immunoglobulins (Igs) that provide the overall disease resistance of the newborn kid. There is virtually no transfer of maternal antibodies across the placenta, so the newborn kid is born with no antibodies and is therefore totally susceptible to disease. Two primary factors are involved: loss of absorptive sites, and bacterial colonization of the intestine. As soon as the kid is born, the ability of the colostral Igs to cross from the intestine into the bloodstream begins to decrease; by twenty-four hours after birth, colostral Ig absorption has ceased. At the same time that colostrum is being absorbed, bacteria are attempting to enter the bloodstream, and so it is vital that the colostrum antibody gets into the blood before the bacteria in order to prevent disease. For maximal protective effect, colostrum should be fed at birth, because with every hour that passes after birth, absorption is decreased; the target for adequate colostral intake is six hours. The timing of colostral feeding is therefore critical.

In addition to its role in transferring antibody, colostrum is also a very rich source of nourishment, being high in protein, fat, and vitamins A, D and E. This readily available nourishment gives protection against environmental insults such as chilling. It also has a laxative effect, suckling stimulating the passage of meconium (the first faeces passed, usually orange/brown in colour, and very thick and 'sticky').

Healthy kids will normally consume sufficient colostrum by suckling. If they can't stand, or if the doe has no colostrum, or there are too many kids to feed, then steps must be taken to ensure that they are given colostrum by artificial means. Although there are many forms of artificial colostrum available, these are no substitute for the doe's own colostrum: this gives a unique 'finger-print' to your goats and your premises, as the colostrum they produce will be as a result of infectious agents they have encountered with you.

Ideally, supplies of fresh colostrum should be available, and this is not usually a problem in larger herds in which a number of does are kidding together. Some units will 'pool' colostrum – literally mixing colostrum from a number of females together, the principle being that good colostrum will counterbalance poorer colostrum. There is one important disadvantage however, and that is the potential for certain diseases potentially spread by colostrum to be widely spread to susceptible kids. One doe infected with either CAE (caprine arthritis encephalitis) described in more detail in Chapter 11, or with Johne's disease (Chapter 7), can contaminate a whole pool of colostrum, with devastating consequences – discuss the status of your goats to these diseases with your veterinary surgeon.

How Much Colostrum to Feed?

Colostrum should be fed at the rate of 50–75ml/kg on three occasions during the first twenty-four hours, with the first feed given in the first hour of life ideally, but certainly within the first six hours. Factors that increase the kid's requirement for heat production, such as bad weather or chilling, increase the demand for colostrum. Housed kids require something like 210ml/kg during the first day, while kids outside in adverse weather may require as much as 280ml/kg. As an example, aim to give a housed kid of a certain weight the following amounts:

These kids are drinking pooled colostrum/milk – beware the spread of disease!

Weight of kid	Amount of colostrum:	
	at first feed	daily
3kg	150–200ml	600ml
4kg	200–300ml	850ml
5kg	250–475ml	1,100ml

Storing Excess Colostrum

Colostrum can be readily frozen, as the antibodies remain stable for at least one year, so you can in theory store a small quantity from one season ready for use at the start of the next season (but not longer). If you have does that are kidding early and are prolific milkers, consider using them as colostrum donors. The easiest way to store colostrum in the deep freeze is in commercially available ice-cube packs; each cube is 20–25ml, thus giving good flexibility when thawing for future use, and avoiding wastage. The packs also lie flat in the freezer, taking up less space. An alternative means of freezer storage is in old yoghurt pots.

To thaw frozen colostrum ready for use takes care and thought, as excessive heat will destroy the vital antibodies – do not be tempted to use a microwave, or thaw over direct heat. You will usually need it quickly, and won't have time to wait for it to thaw naturally, so the best way to obtain it is as follows: place

It is convenient to freeze colostrum in ice-pack containers, then specific amounts can be stored.

it in a bowl, then place the bowl in saucepan containing water, and then heat (this ensures a gradual rise in temperature).

What Value a Neighbour's Spare Goat Colostrum?

This may be a useful alternative in an emergency, but there are two potential disadvantages. Firstly, the colostrum will be fine-tuned to protect against disease in your neighbour's goats, but not yours; thus although the kids will receive the broad nutritional benefits, they may not be fully protected against disease among your goats. The second potential disadvantage has already been covered: if your neighbour's goats have either Johne's disease or CAE, then colostrum may well transfer these two important diseases to your kids, and potentially infect your herd if you are free of infection.

Alternatives to Goat Colostrum

Colostrum from sheep or from cows can be used at approximately the same rate of feeding. It will never be as good as goat colostrum from your own goats, but can nevertheless provide an alternative in an emergency. There have been occasional reports in sheep of lambs developing a severe anaemia and jaundice after being fed cows' milk, due to an adverse reaction. The author has never encountered this in goat kids, but any such suspicion in a kid that is fed cow colostrum should be discussed immediately with your vet before more cow colostrum is used.

There are a number of commercial colostrum replacements (mainly for sheep and cattle), though for reasons stated above (namely lack of specific disease prevention), these should only be considered in an emergency. Read the small print, and ensure that you are confident that it can be fed safely to kids, and that you know what protection it gives – if in doubt consult your veterinary surgeon.

Some shepherds will make up a 'home-made recipe': this will provide nourishment and a laxative effect, but it has no antibody protection and should only be used as a last resort.

The following is a typical recipe:

1.5 pints warm cow's milk
1 beaten egg
1 level dessertspoonful glucose
1 teaspoon cod-liver oil

Feeding a Weakly Kid

So how do I get colostrum into a kid that won't or can't feed? Although you can get a kid to drink from a bottle, there is a danger that it may choke if it is too weak to swallow. A safer and more reliable method is to use a stomach tube.
Never attempt this procedure until you have been shown how to do it, or have at least seen someone carry out the procedure!!

The Technique of Stomach-tubing a Kid
Make sure you use a tube of the correct size; a commercially available lamb stomach tube should be ideal, with a 60ml syringe.

The procedure is best carried out placing the kid over your knee whilst seated, or by laying it on a bale. If you are inexperienced in the procedure, then it may be best to measure the tube along the kid's body, using a pen to mark the length of the tube from the nose to the last rib in order to give you some idea how far the tube has to be inserted. Lubricate the end of the tube, then with your finger in the kid's mouth, and keeping the neck extended, carefully slide the tube down its throat. Don't rush the procedure, and don't force the tube down – the tube should pass fairly easily into the oesophagus, and the kid may aid the procedure by swallowing it.

If you watch the left side of the kid's neck, you should be able to see the end of the tube going down the oesophagus; you can also feel the tube by placing your finger and thumb either side of the neck – it should be distinct from the trachea that can be felt alongside, and has the readily recognizable cartilage rings. Put the end of the tube to your ear, and you should hear a gurgling sound. If the kid becomes distressed, you may have passed the tube into its windpipe (trachea); it will become very agitated, and will look as if it is choking.

When you are satisfied that the tube is in the stomach (and this is where your mark on the tube will help), then – and only then – attempt to pass the colostrum through the

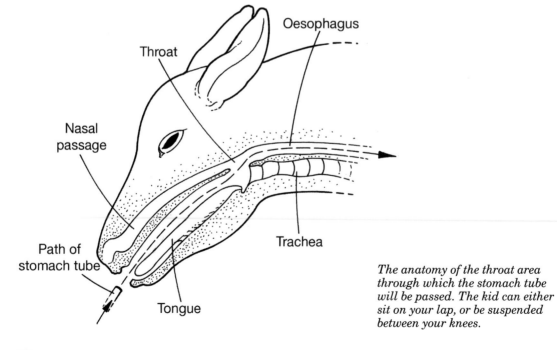

The anatomy of the throat area through which the stomach tube will be passed. The kid can either sit on your lap, or be suspended between your knees.

tube, either by gentle pressure on the plunger of a syringe, or by using a funnel and jug. If the tube is only in the oesophagus, milk will be regurgitated, and the kid may choke.

Gently remove the tube when the procedure is complete.

COMMON CAUSES OF ILLNESS AND DEATH IN YOUNG KIDS

Such problems are often referred to as 'neonatal' – neonatal diseases and neonatal mortality – the term 'neonatal' literally meaning close to birth, and generally referring to problems in the first seven to fourteen days of life.

Hypothermia

When the body temperature of a newborn kid begins to drop, the process is referred to as 'hypothermia', a potentially serious condition that must be recognized and treated or the kid may die. Immediately after birth, when the kid has left the warmth of the womb and is still wet with foetal fluid, it is particularly susceptible to chilling (much like walking out of the sea with a wind blowing even on a warm day). It follows, therefore, that kids born outside are particularly vulnerable, as are kids born indoors but left lying in a draught. There are two critical periods when this problem can develop:

From birth to five hours old: Chilling occurs as described in the previous paragraph, in a kid that has not received its colostrum. The energy reserves will not yet have been depleted, and the blood glucose levels will be within the normal range.

From five hours to three days of age: This is a similar scenario to the above, but these kids will have depleted their energy reserves and become dangerously hypoglycaemic, when blood glucose levels fall (similar to the state that a diabetic can enter).

Not surprisingly, a good stockperson will recognize those vulnerable kids that are most susceptible to chilling, and will take steps to protect them and prevent more cases from developing. Those particularly at risk include 'multiple births' (the kids will have a low birthweight, and there will be possible problems with normal mothering instincts and milk supplies), kids from sick does, kids that had a difficult birth and may have been slow to stand, premature kids, and any kid born outside in inclement weather.

Recognizing Hypothermia

A hypothermic kid will initially look dull and depressed, often standing motionless with a hunched back; this will progress to recumbency, coma and death if its state is not recognized and treated. Other problems can mimic this appearance however, and the only way of confirming your diagnosis of hypothermia is by taking the kid's temperature (*see* Chapter 1); any necessary action depends in part on the temperature recorded:

- 39–40°C (102–104°F) is the 'normal' or acceptable temperature.
- 37.5–39°C (99–102°F) suggests that the kid may be 'at risk'.
- At below 37.5°C (99°F) the kid is hypothermic and needs urgent attention.

Treating Hypothermia

The age of the kid and its body temperature are both important in deciding what to do next, and if you are in any doubt, then seek urgent veterinary attention.

An 'at-risk kid' (37.5–39°C) at any age should be dried, warmed and given colostrum, but kept with its dam and littermates to avoid possible rejection at a later stage.

Any kid with a temperature below 37.5°C (99°F) should be treated as above but you may have to correct the likely hypoglycaemia (*see* below) that will have developed if it is over five hours old, and it will probably need a more intensive approach to save it.

Drying the kid: Physically drying the kid with clean towels should be sufficient, although a source of warmth such as an infrared lamp will also help to dry it. Don't be tempted to use a hairdryer – although the coat will invariably dry out, you may create a steam made up of water droplets that could contain potentially harmful bacteria that you could inhale (the author is aware of at least one Q Fever incident in which this occurred, when a family developed the illness after weak and infected lambs were brought into the kitchen and dried with a hairdryer).

Glucose injection: Do not attempt this yourself unless you have been shown how to do it, preferably by your own veterinary surgeon, or have discussed the procedure with someone who has carried it out. Many sheep owners are familiar with the technique. Remember the procedure is necessary for kids more than five hours old that have been shown to be hypothermic.

- Your calculated dose rate is 10ml/kg of a 20 per cent glucose solution. Most proprietary glucose solutions are purchased as 40 per cent solutions, and you will need to dilute each ml of 40 per cent solution with an equal amount of either sterile water or boiled (and cooled) water to achieve this. Whatever approach you adopt, warm the solution before injecting it.
- Suspend the kid by its front legs with its hind limbs hanging down.
- Locate the navel stump, and with either an antiseptic or antibiotic aerosol spray, cover an area 1cm (3/8in) to the side and 2cm (3/4in) below the navel, as your injection site.
- Attach a sterile 19g 25mm needle to your measured dose of warmed solution in the syringe, and push the needle into your target site with the needle at an angle of 45° to the skin, directed towards the hip area.

The warm solution will quickly flow around the intestines and will be rapidly absorbed. Some lambs when injected may urinate, so don't worry if this also happens with your kid; it appears to be a reaction to the warm fluid and possibly to the discomfort of the needle passing through the skin. The result can often be quite spectacular!

Warming: It is all too easy to overdo things

when you start trying to heat the lamb up – the temperature gradient is very narrow, so that a lamb may die from hypothermia – being too cold – at 37°C, but from *hyperthermia* if the body temperature rises above 40°C, a safety margin of only 3°C! There are a number of options available. These include:

- Bringing the lamb indoors in front of the kitchen range or gas fire, though remember the zoonotic risks described earlier.
- Infrared lamps – these are widely used on many farms, but must be used carefully to ensure that the kid is not burnt or allowed to overheat. Cardboard boxes are useful, as they can be burnt after use, thus minimizing the chance of disease being spread. Suspend the lamp at least 1.2m (4ft) above the box or penned area.
- Warming boxes – these can be purpose built or made as a temporary measure with straw bales. Warm air is blown in with a fan heater – great care must be taken to avoid the risk of fire, however, and careful monitoring is required.

With any of these options, kids should be monitored and their temperature taken at least every 30min. Once the temperature has reached 37.5°C, they can safely be removed, and preferably placed back with the dam.

Diarrhoea

Diarrhoea (or 'scour') is a relatively common problem in kids during their first week of life, when they are particularly vulnerable; kids that have received insufficient colostrum are more at risk. Diarrhoea can be caused by either dietary factors or a number of infectious causes.

Dietary Scour

This occurs occasionally in artificially reared kids on a milk substitute. Provided the feeding is regular and consistent, however, this is the most cost-effective way of feeding kids, with problems only developing when any inconsistency creeps in; a sudden change from whole milk to milk substitute and vice versa, changes in the type of milk replacer fed, overfeeding, or varying the temperature and concentration, can all precipitate problems. Problems can also result from dirty utensils or other milk feeding equipment, or from contamination or spoilage of the milk substitute.

Infectious Scour

A number of infectious agents have been incriminated as causing scour. Your veterinary surgeon may ask you to collect a faeces sample to submit for laboratory examination. If you haven't got a laboratory sampling pot, use an old (but clean) yoghurt pot, and try and collect samples from untreated kids if possible. Once your veterinary surgeon has identified the underlying cause, then suitable control measures can be put in place to treat any further cases, and to prevent new cases developing. Treatment with electrolytes is important to control the dehydration and electrolyte loss that occurs, and it is important to avoid overcrowding. Many kids that recover may remain stunted due to damage to the gut villi.

Avoid overcrowding where possible, as infectious agents (particularly those causing scour) can spread rapidly.

E. coli: Although the normal gut of a kid will contain a population of *E. coli*, most are harmless and involved with the normal digestive processes. There are, however, some strains, referred to as enterotoxigenic or enteropathogenic, that have the capability of producing powerful toxins and adhering to the gut wall, causing severe damage as a result. The toxins cause acute and profuse watery diarrhoea, rapidly resulting in dehydration and death. Your veterinary surgeon will prescribe antibiotics, but other supportive treatment, particularly fluid therapy, is required to reverse the toxaemia.

Salmonella infection: This is a comparatively rare problem in goats, compared to other farmed species such as cattle. As the organisms that most commonly cause infection (*S. typhimurium*, *S. enteritidis*) are widely distributed in farmed and wild animals and birds, however, the potential for cross-infection is always a possibility. Possible entry routes for infection include wild bird/ rat contamination of feed or bedding, indirect transfer of faeces on wellington boots, or borrowed equipment from other infected farm animals, for example cattle. Healthy carrier goats may also act as a source of infection. This is another potentially fatal diarrhoea, and veterinary attention should be sought as a matter of urgency.

Signs again include profuse watery diarrhoea, often with blood and portions of 'shredded gut lining', abdominal pain, pyrexia, dehydration and death.

This organism is a real risk to human health (a zoonotic organism), and steps should be taken to protect the health of all humans in contact with the kids (*see* Chapter 17).

Cryptosporidia: This is a protozoal parasite that multiplies in the gut wall causing malabsorption and profuse watery diarrhoea. Most cases occur in kids more than one week old (*see* Chapter 7). Yet again, this infection poses a risk to human health, and in particular to children playing with, or bottle feeding very young goat kids with or without diarrhoea.

Gut in cross-section

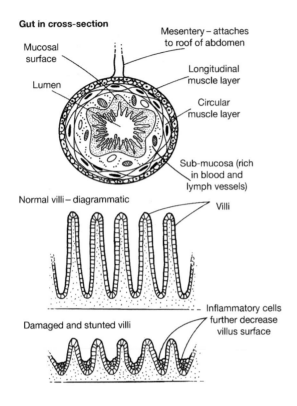

The gut in diagrammatic cross-section; note the villus structure increasing the surface area.

Miscellaneous Neonatal Problems

Border Disease (BD)

Although more commonly encountered in sheep, this virus has also been reported in goat kids. BD is caused by a 'Pestivirus', a group of viruses that have the ability to cross the placenta of a pregnant susceptible animal, and damage the nervous system in developing foetuses. Some kids will be aborted, and some will be born alive, apparently fit and healthy, but carrying the virus. A proportion of infected kids will, however, be born alive but showing a range of neurological signs, from a fine body tremor to severe brain damage.

Brain Damage/Hypoxia

This can occur in any kid that has had a prolonged or traumatic birth, causing the brain to become starved of oxygen. It is most commonly associated with backwards/breech presentations, or with a large kid when only the head is presented and becomes stuck. Affected kids may appear disorientated when born, but normally improve quite rapidly, providing they are given colostrum by stomach tube and kept warm. If there is no real improvement after twenty-four hours, then euthanasia on humane grounds may be a kinder option.

Congenital Abnormalities

Such conditions are apparent at birth, being the result of an abnormal foetal development that may or may not also have a hereditary basis.

Supernumerary and abnormal teats: Anything more than two teats is abnormal. Additional teats may be apparent at birth, but may not be visible until a few weeks of age. Occasionally two teats may be joined at the base, or the tip may be abnormal – known as a 'fish-tail' teat. Most are innocuous and remain blind; others may develop mammary tissue at a later stage. In the show world, judges will normally disqualify a goat with extra teats, and for this purpose (and because they are often hereditary) it is probably unethical to remove them. Simple extra teats are, however, often removed in commercial herds (usually at the same time the kid is disbudded) due to the potential for them to become infected. Discuss the best approach with your veterinary surgeon.

Overshot/undershot jaw: Although comparatively rare, this is reportedly more common in the Anglo Nubian breed. Unless so severe that normal feeding behaviour is prevented, most kids can be successfully reared.

Entropion: This is a potentially severe condition that must be recognized quickly, before permanent eye damage occurs. The eyelid is turned inwards, so that the lashes rub on the cornea; it is the lower eyelids that are most commonly affected. The lid can be 'rolled' out, but will usually return. Your veterinary surgeon will normally carry out a minor surgical procedure to correct it.

Intersex/hermaphrodite: Severe cases can be recognized due to gross abnormality of the external genital organs; others may not be

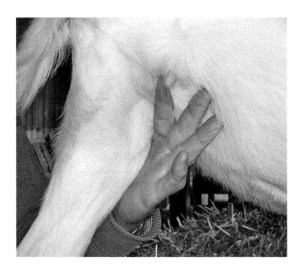

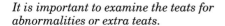

It is important to examine the teats for abnormalities or extra teats.

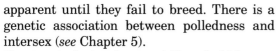

This is referred to as a 'fish-tail teat'.

apparent until they fail to breed. There is a genetic association between polledness and intersex (*see* Chapter 5).

Contracted tendons: Affected kids are unable to straighten their front legs properly. They may stand on the tips of their toes (and most of these will recover spontaneously with time), or the legs will be bent to such an extent that the kid may walk on the fronts of its fetlocks. With the latter presentation there is a danger that the skin surface will be broken and become infected. Padding such limbs with cotton wool and a splint may help, but ensure that the splint is not applied too tightly. The condition may be related to lack of space in the uterus as the foetus develops.

Over-extension: There are occasional reports of an apparent over-extension of hock and carpal joints, but again, this will generally correct itself. There may be a predisposition in the Boer goat breed.

Wryneck: The author has encountered one incident in which this condition was widespread among a group of closely related Boer goats. In wryneck (also referred to as torticollis), the kid is born with a twisted or distorted neck, usually with the head drawn to one side by rigid contraction of the muscles, which cannot be straightened. An investiga-

Controlling Abnormalities
Always record the identification of any doe that produces abnormal kids as above. If they produce the same kids in successive years, particularly with the same matings, then the condition may well be hereditary. Owners of pet goats may decide not to breed from them again, and owners of commercial goats may decide to cull.

tion traced all offspring back to a common ancestor, and attempts have been made to breed the condition out by selective matings.

Difficulty passing faeces:
Retained meconium This is the first faeces passed by the newborn kid, and is produced while the kid is developing in the womb. Colostrum acts as a laxative, and if the kid does not have a good early feed it can be retained. In addition to blocking the normal flow of faeces when the kid begins to feed, it is also an indication that colostrum has not been consumed, and the kid may be especially vulnerable. If the kid is straining or obviously distressed, then you can very simply give it an enema by passing 15–20ml of warm soapy water through an old stomach tube (not the same one you use for tubing the kids with

This kid is affected with wryneck/torticollis and is unable to straighten its neck.

colostrum!) gently inserted about 5cm (2in) into the rectum. The kid will begin to strain to expel the warm water, and hopefully the meconium will follow, although you may have to repeat the procedure.

Imperforate anus More correctly termed 'atresia ani', in which there is no opening from the rectum through which faeces can pass. You may notice this when the kid is born – particularly if you attempt to take its temperature! Other cases, however, are not recognized until the abdomen begins to swell, and the kid begins to look uncomfortable and strains constantly as if trying to pass a motion. This is an emergency, and if untreated will be fatal, so contact your veterinary surgeon. Those that respond best to treatment are those where the anus is closed off simply by a thin layer of skin; a single cut under local anaesthetic will release a mass of foul-smelling faeces, and the kid will be much more comfortable almost immediately. In more complicated cases, however, there may be an absence of part of the rectum, from a few millimetres to several centimetres

in a grossly deformed gut, and the success rate in these is much less predictable.

Floppy kid syndrome: This condition appears to be becoming more widespread in the USA, having been first recognized and reported as recently as 1987. It does not appear to have been reported in the UK, to the author's knowledge. The condition presents as a sudden onset of profound weakness, with an inability to stand or hold their head up, in kids aged 3–10 days, that are apparently normal at birth. It appears to be related to a metabolic acidosis (usually associated with neonatal diarrhoea), but in this condition, no other organ systems are involved. Response to treatment with intravenous fluids and supportive care can be dramatic, although outbreaks in herds have reported as many as 50 per cent of kids affected in a season, with the case fatality as high as 50 per cent! Diagnosis is based on the typical clinical signs described, the elimination of other causes of acidosis, and the response to treatment once recognized.

Fractured jaw or limbs: Dystocia (a difficult kidding) is not common, and fortunately injuries to the kid during delivery are comparatively rare. Excessive force, particularly when applied by an inexperienced stockman, can result in a fractured or dislocated jaw, or a fractured limb, as both are very easily damaged. Limb fractures can also occur if kids are trodden on or kicked, particularly by inexperienced does. Long bone fractures will heal very quickly, and most veterinary surgeons will treat these by applying an external cast.

Goitre/iodine deficiency: Although young goat kids are occasionally presented with a swelling in the neck, and an enlarged thyroid gland or 'goitre' is suspected, true iodine deficiency in goats in the UK is rare, and in fact many of these throat swellings involve not the thyroid gland, but the thymus. The age of the kid gives a clue, since thyroid swellings are usually more apparent in kids up to two to three weeks of age, whereas thymus gland swellings are usually more apparent over three weeks of age (Chapter 9). In true iodine deficiency, kids are usually stillborn, or born weak and sickly, being susceptible to the cold and often with a poor, sparse, hairy covering. The thyroid gland itself normally weighs around 2–4g, but may increase in size up to 50g in response to low iodine intake, or to certain feedstuffs referred to as goitrogenic plants – for example, the brassica family – blocking its uptake. Once confirmed, then iodine supplementation must be an annual procedure during pregnancy.

Hernias: From time to time, kids will be born with a weakness around the navel; these can either be closed or open hernias. In a closed hernia, a swelling can be identified around and above the stump; this can be corrected by applying gentle upward pressure, when the swelling disappears back into the abdomen, and a 'ring' can be recognized through which the content of the swelling (usually a length of intestine) has passed through. These are best left, as many will disappear spontaneously when the kid grows. An open hernia is another genuine emergency, however, as in these kids (usually identified very shortly after birth) the protective layer of skin covering the abnormally developed abdominal wall around the navel has also failed to develop. Intestine, or even the abomasum, can prolapse straight through this hole. The kid quickly becomes very shocked, and there is the risk that the exposed gut can be trodden on by other goats, or become heavily contaminated with bedding and faeces. Your veterinary surgeon may decide to attempt to repair the damage surgically, but the success will depend on how long the gut has been exposed and dried out, and how much damage and faecal contamination has occurred. Phone your veterinary surgeon for immediate advice, but gently pick up the kid, supporting the prolapsed gut in a clean and moist cloth or towel, and keep the kid warm to control the shock and prevent any further damage occurring. The procedure to repair the hernia may be best carried out at the surgery. If there is severe damage to the gut, then euthanasia on humane grounds is the only course of action.

Joint ill, navel ill, spinal abscesses and liver abscesses: This group of problems is more commonly encountered in goat herds kidding indoors, particularly if the kidding area has been in use for some time, and is becoming either contaminated, or overcrowded, or both. Infection usually with *E. coli* or *Streptococci* appears to gain access to the body through the navel, particularly if it is slow to heal and/or has not been dipped. In navel ill, the navel feels very thickened and is painful, and resulting infection tracks either via the umbilical vessels to the liver, causing large abscesses to develop – 'hepatic necrobacillosis' – or can lead to a severe peritonitis, both of which can be fatal. (For joint ill *see* Chapter 11, and for spinal abscesses *see* Chapter 10.)

Swayback: *See* Chapter 10.

White muscle disease: Although losses related to this condition are most commonly seen in older kids, severely affected newborn kids may die in the first few days of life due to damage to cardiac muscle. A common presentation is 'sudden death' literally from a heart attack due to cardiac muscle damage. If found in the early stages of disease, kids are usually dull and depressed, and have a fast heart rate and difficulty breathing as heart failure sets in. The condition is related to a deficiency of selenium/vitamin E. Your vet will be able

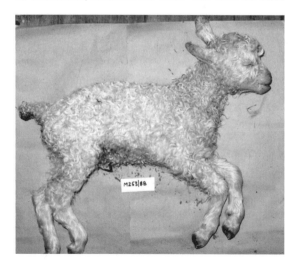

This highly valuable imported pedigree angora kid died as a result of navel ill and septicaemia. It was moved to a pen that had previously housed lambs and had not been cleaned out.

to distinguish this disease from others such as joint ill that may show similar signs. *See* Chapter 11.

DISBUDDING

This is the procedure whereby the horn buds are removed before any horn material has really had a chance to develop. The horn buds of goat kids can grow very rapidly, and it is recommended that if carried out, the procedure should be completed in the first week of life, preferably between days two and seven.

It is important to stress that disbudding is a mutilation, and should not be carried out purely as a routine procedure. Many pet goats such as Pygmy goats kept singly, for example, may not need to be disbudded. The procedure is much less stressful than removing the horns of an adult, however (as illustrated). Whether to disbud or not is an important decision to make, so ensure that you talk this through with your vet, remembering the short time-

scale in which disbudding can be carried out. There is no doubt that horned goats can be a danger to other goats, and also to themselves as they are more prone to getting caught in fencing, haynets and so on. Never mix horned and polled goats, as the horned goats will quickly realize that they have an advantage and may bully others. Remember also that a large goat with horns (particularly a mature male) may be very difficult to handle.

In the UK, a goat kid can only be disbudded by a veterinary surgeon, and the kid must be anaesthetized. Your veterinary surgeon may use a local anaesthetic nerve block, and there are two nerves that require blocking to each horn bud. The dose of local anaesthetic necessary in a very small kid can, however, approach the toxic limit, and for this reason the majority of kids are disbudded under a general anaesthetic, using either an injectable or gaseous anaesthetic. Many vets prefer kids to be taken to the surgery, although for practical reasons, larger numbers can be disbudded on the farm.

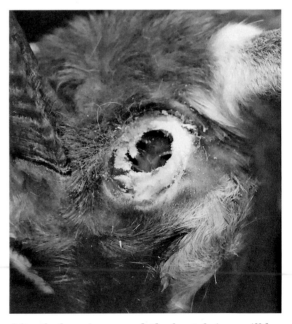

After a local anaesthetic has been given, the horns are removed in this instance by using a length of 'cutting' wire.

After the horn is removed, the frontal sinus will be exposed, leaving a large hole that will eventually cover over.

The goat kid's skull is particularly thin, and great care must be exercised to ensure that the bud is destroyed, but that the underlying brain is not damaged; for this reason a very hot iron is used, so it need be in contact with the head for only a minimal amount of time, thus minimizing the chance of heat transfer through the thin skull. The damage that can be done in unskilled hands is enormous (as illustrated in colour section).

Kids that have been disbudded are often 'off-colour' for a few days, and they should be carefully observed to ensure that they continue to feed. Any sign of nervous disability such as a head tilt, a 'star-gazing' attitude, or convulsions or fits may indicate a complication such as traumatic injury to the underlying brain, or secondary infection of the disbud wound, and your vet should be contacted without delay.

CASTRATION

It is recommended that uncastrated (entire) male kids should not be reared, unless either they are required for future breeding, or if they are to be killed at a young age for meat. An uncastrated male would certainly not make a good pet, as they are likely when mature to be aggressive and very 'smelly'!

As a result, therefore, not many goat kids require castration in comparison with, for example, lambs, in which it is a routine procedure in most lambs born and reared for meat. The preferred method is similar to the procedure used in lambs, however, namely the rubber ring or elastrator ring applied in the first week of life. In the UK, the law states that the procedure can be carried out by an unqualified person aged eighteen or over (or seventeen if undergoing training by a vet or at a recognized training institution such as an agricultural college), and no anaesthetic is required. If you have never carried out this procedure previously, then instruction should be sought from your veterinary surgeon, or from an experienced goat owner or a sheep farmer before you attempt it.

It is advisable to ensure that the kid is protected against the risk of tetanus developing (if the dam was vaccinated against the clostridial diseases, and the kid consumed sufficient colostrum, it should be covered) before you proceed, although remember that seven days is the maximum age for castration by this means.

How to apply the ring:

1. Hold the kid by its forelegs with its tail hanging down, and rest him against your knee; this way you are less likely to trap other organs such as the penis or teats.
2. Check that both testicles are present in the scrotum.
3. Place a ring on the applicator; place it over the scrotum at its base and then release.
4. Re-check to make sure both testicles are below the ring, if you are unsure, remove the ring and re-apply.

Keep the kids in a clean environment, and in time the scrotum and testicles will shrivel and drop off; there is no need to worry about removing the ring at a later date.

It is inevitable that the procedure produces some discomfort, and you may notice kids showing varying degrees of pain for an hour or two after the procedure has been completed. If discomfort continues for a longer period of time, then seek veterinary advice; one possible complication is a trapped penis, thus preventing the kid from urinating.

Castrating an older male goat: This must be undertaken surgically by a veterinary surgeon.

UNWANTED MALE KIDS

On many dairy goat units there are inevitably male kids born that are unfortunately of no economic value, unless a local meat market has been identified, and currently the demand for kid meat is not high. Some may be kept for future breeding, but the majority are destroyed at, or close to, birth on economic grounds.

It is important that the welfare of these kids is not compromised, however, if there is any delay before euthanasia; thus they should receive the full care that other kids receive, including colostrum, warmth and feeding.

All goat farms should have an effective procedure in place to deal with unwanted kids as quickly and humanely as possible; to this end the Humane Slaughter Association in the UK has produced an advisory leaflet (number 7). The association suggests four main alternatives:

- By a veterinary surgeon using intravenous barbiturate.
- By a heavy blow to the head followed by bleeding out by severance of the main blood vessels.
- By the use of a free bullet.
- By the use of a captive bolt stunner followed by pithing or bleeding out.

None of the above procedures should be carried out by anyone who is not confident and competent enough to carry out the task humanely.

The farm must also have a policy for the disposal of these kids after euthanasia.

CHAPTER SEVEN

The Digestive System

Feeding and digestion is a continual process in goats and other ruminants, and you will notice that they spend much of their day either eating or ruminating. When these procedures stop or are reduced in frequency, it may be an early sign of illness. Although a reduction in appetite may be an indication of a digestive upset, it is also a vague general indication that the goat may be sick, for a wide variety of reasons including, for example, mastitis, pneumonia or severe lameness.

PROBLEMS AFFECTING THE MOUTH, TEETHAND OESOPHAGUS

The diagram on p. 84 shows the pattern of dentition in the goat, with times of tooth eruption; it is adapted from Owen, 1977 (*The Illustrated Standard of the Dairy Goat*, Dairy Goat Journal Publishing Corporation, Scottsdale, Arizona).

Goats only have incisors on the bottom jaw; these are in direct contact with a dental pad on the upper jaw. Molar and premolar teeth grow constantly, and anatomically the upper and lower jaw teeth should be in apposition with each other, ensuring a constant and consistent wear as the biting surfaces grind together.

As goats age, however, there is increasing wear of the incisor and molar teeth, such that normal feeding behaviour may become more difficult. Real problems can develop when there is uneven wear, or a cheek tooth is lost through gum disease, such that the opposing tooth has nothing to grind against. Under these circumstances, sharp projections can develop, often resulting in mouth sores and ulcers. Owners should check for uneven tooth wear, but remember not to put your fingers in the goat's mouth – the teeth are very sharp, and can cause real damage to your fingers!

Quidding

If tooth wear becomes a problem, then goats will often be seen to drop food from their mouth as they are chewing, a condition referred to as 'quidding'. Other causes of quidding might include the following:

Foreign bodies: For instance thorns, pieces of wood (often wedged between the teeth), string or wire wrapped around the base of the tongue – remember goats are inquisitive!

Drenching gun injuries: If a goat has received a dose of, for example, a wormer in the past few days, it is possible that damage has been caused to the pharynx behind the base of the tongue.

Stomatitis: Literally this means a 'sore mouth' – the goat's inquisitive nature may lead it to lick something potentially caustic or hot in its environment. A plant known as 'giant hogweed' can present a further hazard, as leaking juice from its cut stem (after mowing or strimming) can lead to blisters developing

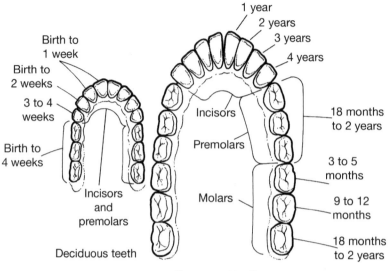

Birth to
1 week

Birth to
2 weeks

3 to 4
weeks

Birth to
4 weeks

Incisors
and
premolars

Deciduous teeth

1 year
2 years
3 years
4 years

Incisors

Premolars

Molars

18 months
to 2 years

3 to 5
months

9 to 12
months

18 months
to 2 years

Permanent teeth   *Goat dentition.*

on the lips, tongue and gums. It is a particular problem in those goats with a white muzzle and white mouth rather than the pigmented goats, due to the interaction with sunlight, and the protective benefit of skin pigment.

Orf lesions: These can cause young kids to dribble milk when feeding (*see* Chapter 12).

Vomiting

Although goats in theory cannot vomit, there is one situation in which they can appear to project a cud from their mouth as if vomiting. This follows rhododendron poisoning, a plant that may be eaten by the inquisitive goat, particularly if there is little else to eat (*see* Chapter 16).

Choking

Goats can occasionally get a piece of food material caught at the back of the throat, or during its passage down the oesophagus; examples include portions of apple or carrot. If it is at the back of the throat, the goat will be very distressed and will salivate profusely. You may be able to see the obstruction yourself and remove it (remembering not to get bitten), but contact your veterinary surgeon urgently if it can't be dislodged; your vet may need to administer a sedative to reach

it. If the obstruction has passed through the pharynx and is now lodged in the oesophagus, the goat will be less distressed initially, but it will be unable to eructate (belch) gas from its rumen, and will rapidly become bloated – yet another veterinary emergency (*see* later in this chapter).

PROBLEMS AFFECTING THE FORE-STOMACHS

The fore-stomachs of the goat are called the rumen, reticulum, omasum and abomasum. They occupy a considerable amount of space within the abdominal cavity, as indicated in the diagram opposite.

Bloat

This term refers to the accumulation of gas causing the rumen to become abnormally distended. Most commonly the gas accumulates in the rumen, when the swelling first becomes apparent in the upper part of the left flank where there is normally a natural 'hollow'. Gas can also accumulate in the abomasum (almost always in artificially reared young kids), and in the intestines as a result of an obstruction such as may occur in a 'twisted gut'.

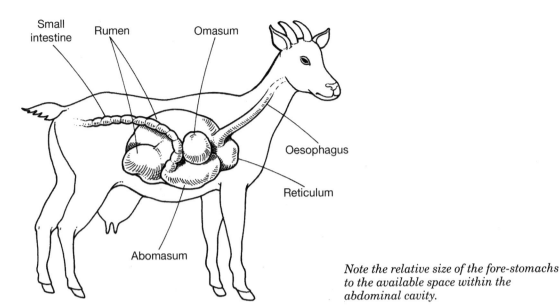

Small intestine — Rumen — Omasum — Oesophagus — Reticulum — Abomasum

Note the relative size of the fore-stomachs to the available space within the abdominal cavity.

Bloat can be a life-threatening problem, it may develop very quickly, and the goat will become distressed as a result of pressure on its diaphragm that makes it difficult for it to breathe.

Bloat in Adults

The sudden appearance of bloat in an adult is most likely due to one of two problems: the goat has been grazing lush herbage, and particularly clover or lucerne (although the normal browsing goat is unlikely to be affected), or it has been supplied with grass cuttings that have fermented in the rumen; this type of bloat is generally referred to as a 'frothy' bloat, as the gas bubbles are trapped in the mass of fermenting vegetable material literally as foam, and it is always an acute emergency. The second form of bloat (referred to as obstructive bloat) differs, in that the gas is trapped as a cap over the surface of the normal rumen content, mainly due to an obstruction to the belching mechanism that accompanies rumination. This may be acute and can occur in, for example, choke (*see* earlier in this chapter); or it can be chronic, when the bloat develops, then is released as pressure builds by normal belching, but then recurs, gradually increasing in frequency and/or severity. The main causes of chronic bloat are abscesses or tumours developing along the length of the oesophagus and causing pressure, thus preventing the normal passage of gas.

Urgent veterinary attention should be sought, particularly if the goat is very distressed. Depending on the cause, your vet may pass a stomach tube to release the gas, or may even push a needle or trochar through the skin and underlying rumen wall to release it. Do not attempt either of these procedures yourself, unless you have been shown how to, because you may cause more harm than good. If there is to be a delay in treatment, then ask for veterinary guidance on what you should do. Frothy bloat cases may benefit from drenching with a bloat drench (giving the sheep dose), or with 100–200ml of vegetable oil; or in an emergency 10ml of washing-up liquid can be administered – gently massage the abdomen after drenching in order to distribute the liquid. Standing the goat with its front end raised above its back end will help it to breathe more easily.

Bloat in Young Kids

Young artificially reared kids are quite susceptible to bloat; this is normally related to incorrect or unsuitable feeding methods. Problems may occur as a result of:

- Feeding too fast, often due to a teat orifice that is too large or to a damaged teat end. Milk will not follow the normal groove directing it to the abomasum, but will overflow into the rumen and ferment.
- Inconsistent feeding – kids need to be fed the same quantity of milk, at the same temperature and concentration and at the same time each day. Any deviation from this can lead to problems such as bloat.

Most cases tend to be chronic and insidious, and are best treated by early weaning if possible. As with adults, however, any sudden bloating in a young kid is an emergency requiring veterinary attention or advice. It is a cause of 'sudden death' in kids, but a post mortem is necessary to confirm this, as many kids that have died for other reasons will develop bloat after death.

Over-eating (acidosis)

Goats must eat constantly for the rumen to function, and the good stock-keeper will ensure that they have a good and varied diet. A sudden change in diet, particularly eating a lot of highly fermentable food such as concentrates or cereal (most commonly when the feed-store door is left open, or a bag of feed is left supposedly out of reach) is, however, potentially fatal. The goat will rapidly eat what is often very palatable food, but problems can develop almost immediately in the rumen, where harmful amounts of acid and gas are produced. The normal rumen micro-organisms are killed off, and a serious digestive upset is the result.

Mild cases will be 'off colour' for a day or so, may scour, but will usually recover. Severe cases may be found dead within hours of gorging themselves, the rumen acid that is produced having a profound effect on the circulatory system. Less severe cases will look very dull and depressed, and will often be bloated and scouring. Veterinary attention should be sought, but there are many complications that can develop, and the success rate

may be disappointing.

The moral is to ensure that all food stores are goat proof, and to introduce new constituents to the diet gradually.

Rumen Indigestion

Having described the two extremes of rumen bloat and acidosis, it is perhaps important to stress that the complexity of the rumen digestion is such that many minor 'indigestion'-like problems will occur. These are usually vague, and include inappetance, mild rumen bloat that eases itself, pasty faeces, and depressed milk yield. Many of these upsets are self-limiting, but the good stockperson will learn to recognize when things are potentially more serious.

PARASITES OF THE STOMACH AND INTESTINE

Worms – parasitic gastro-enteritis (PGE)

Parasitism caused by nematode worms in the stomach and gut is probably the most important cause of production loss and even death in goats reared together while grazing, and can affect any type of goat, from those kept singly to larger groups. It is the most significant reason why many commercial goat herds are kept indoors all year round, due to the practical and economically expensive need to control worms. By housing, and removal from grass, the risk of picking up a worm burden is removed.

How do goats get worms?
A worm burden is picked up when goats graze pasture contaminated with worm larvae. Once the larvae reach their preferred region of the gut, they develop into adult worms, each of which is capable of laying hundreds of eggs into the gut, which then pass out in the dung. Under suitable conditions, larvae hatch from the eggs and are soon ready to infect further goats.

Do all goats get worms?

Yes – this is part of the problem! The effect of the burden can vary from goat to goat depending on the goat's age, the number of worm larvae taken in, and the previous exposure history (i.e. immunity). Young goats in their first year of life are particularly susceptible, and older goats will develop some immunity, but *this immunity is poor and short-lived*. It is vitally important that goat owners reading this book are aware of this – adult goats must be included in any worm control programme established (unlike other ruminants such as cattle and sheep in which adults become immune after exposure in their early grazing life).

When do goats get worms?

The actual intake of worms depends for the most part on the concentration of infective larvae on the pasture, and this can vary throughout the year. Survival and availability of larvae is greatest when the weather is warm and moist, and conversely larvae are eventually destroyed during prolonged dry, cold or hot weather conditions.

The danger period, therefore, may extend from spring to early winter, with the greatest risk of infection in the autumn.

Can sheep or cattle be infected?

As goats, cattle and sheep are all ruminants, many of their diseases are common, and this has been referred to throughout this book. It is particularly true of parasitic diseases, to the extent that any control programme for one species must include any other ruminant species that is co-grazing. There is much evidence to suggest that goats are more susceptible to picking up worms than both sheep and cattle and, more importantly, are at a greater risk of developing gut damage, illness and death than other ruminants. To put this in context, work carried out in Australia compared stocking rates of goats with merino sheep, such that when heavily stocked at the same stocking rate, at slaughter there were sixty-six times more worms in the gut of the

goats than in the gut of the sheep. Much of this is related to the greater ability of sheep to produce immunity to the worm burden than goats, as already mentioned.

Why is the immunity produced against worms so poor?

This may be related in part to the way that goats have evolved and adapted to their environment compared with sheep. Sheep in the wild will graze at or near ground level, where they are likely to pick up nematode larvae. Goats, by contrast, are browsing animals, and in the wild will eat grass and ground vegetation, but will mix this with shrubs, trees, hedgerows and so on, which are more distant from the larvae, hence fewer will be consumed.

Modern husbandry systems have over-ridden this, however, particularly when stocking rates go up, and fencing is used to focus goat activity away from trees and hedges on to a patch of grass that quickly becomes contaminated. In effect we have converted a browsing animal that is adapted to taking in low levels of worms to a grazing animal unaccustomed (through its evolution) to coping with a heavy worm challenge.

What are the main clinical signs to look out for?

- Reduced growth rate and weight loss
- Reduced milk production
- Reduced or poor quality fibre growth
- Diarrhoea
- Anaemia
- Sudden death

The developing larvae in the gut and stomach cause severe damage to the lining, causing loss of protein from the blood vessels into the intestines, and also damaging the cells that are involved in the digestion and absorption of food. The end result is poor appetite, loss of condition, diarrhoea and dehydration.

Even when treated, however, the damage to the villi of the gut can be so severe that

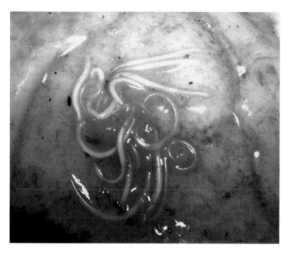

Worms visible on the lining of the gut.

recovery may be slow. And if a new burden is picked up during this 'recovery phase', then the situation will worsen, and permanent gut damage and stunting may be the result.

What worms are involved?
The following worm species will be found in the fore-stomachs of the goat:

Abomasum: *Haemonchus contortus* ('Barbers Pole worm'), *Ostertagia* (now referred to as *Teladorsagia*), *Trichostrongylus axei*.
Small intestine: *Trichostrongylus* species, *Nematodirus* species.
Large intestine: *Oesophagostomum columbianum*, *Chabertia ovina*, *Trichuris ovis*.

The main species causing diarrhoea are *Ostertagia* and *Trichostrongylus*.

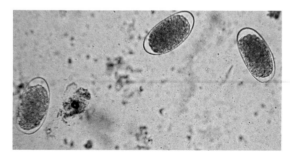

Worm eggs under the microscope; note the other debris present.

Haemonchus contortus is associated with severe wasting and anaemia, usually without diarrhoea, and can be easily overlooked – most stock-keepers associate worms with diarrhoea.

How can I tell if my goat has worms?
Discuss this with your veterinary surgeon. In the live goat, one very useful approach is to examine faeces' samples to look for the worm eggs that are passed out. Any veterinary laboratory will be able to undertake this test, and there are also kits available to carry out the procedure on the kitchen table.

A worm egg count undertaken on a faecal sample will give some indication of the type of worm involved, and the severity of the infection. Care must be taken with interpretation however, because while some species of worms produce high numbers of eggs, others are less prolific. Equally, to demonstrate worm eggs in the faeces relies on the goat having an adult worm burden in the stomach or gut. Severe disease can result from heavy larval challenge before the worms reach maturity, and before worm eggs will appear in faeces – disease caused by *Nematodirus* is a classic example of this phenomenon, and kids can die from acute *Nematodirus* infestation without a single egg appearing in the dung.

The period of time between larvae being taken in by mouth and eggs appearing in the faeces is referred to as the pre-patent period, and this varies between eighteen and twenty-one days for the worms described.

How can I control and prevent worm problems?
As already stated, the larger commercial goat-herds have recognized that production is often unsustainable when goats are kept at grass, and have taken the decision to house their goats throughout the year. If goats have no access to grass, and are housed in this way, and were free of worms when housed, then they will remain free of worms.

Any goat spending some, or all of its time grazing or browsing outdoors, conversely will

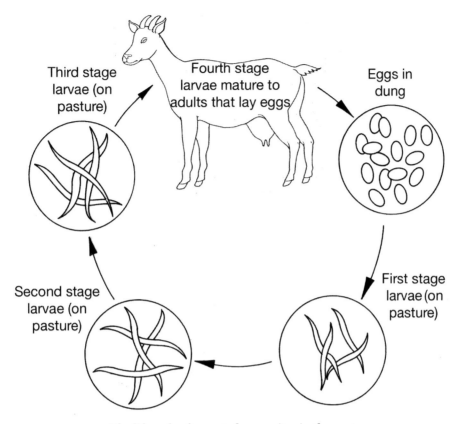

The life cycle of nematode parasites in the goat.

be at risk. The risk itself is difficult to quantify, but as a hypothetical example, a single goat reared indoors, and turned out on to land that has never had goats or any other livestock grazing on it, is likely to remain free of worms indefinitely. This single goat could, however, become infected by the following means:

- By the introduction of another goat carrying a worm burden that would result in eggs being passed on to the clean grazing.
- By the introduction of infected sheep or cattle on to the grazing (remember many worms are shared), thus again resulting in egg excretion on to the pasture.
- By the goat moving to another area of grazing that is contaminated with worm eggs (and this may have occurred for management reasons, for instance there

was no grazing available, or the field was waterlogged), or the goat may have escaped to a neighbour's field. Maintain those boundaries!

With these broad principles in mind, it follows that control can be achieved by two main strategies:

1. Grazing management.
2. Strategic dosing with a wormer (anthelmintic).

Grazing management: A paddock can only be considered to be safe from worms if goats, sheep or cattle have not grazed it for at least twelve months. If silage or hay has been taken from a field, then the resulting aftermath is likely to be fairly clean, assuming that it was not heavily grazed during the springtime.

By housing, large commercial herds remove the need to control worms.

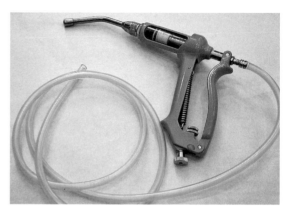

A drenching gun is a useful way to administer a wormer, but must be calibrated and maintained regularly.

There is a particular problem with *Nematodirus*, as the eggs are very resistant, and can readily survive from one spring to the next. For this reason, it is better to avoid grazing kids on the same pasture grazed by either kids or lambs the previous spring.

There is never a time when potentially contaminated pasture is totally safe, and goat owners should be careful about turning goats out in winter on a fine day. They could acquire new infection from the overwintered larvae, and/or could seed the pasture with eggs. Some of these eggs would survive to slowly begin their development in spring, causing a rise in larval numbers on the pasture in mid-summer.

There may be some success in delaying the turnout of kids until they are bigger and stronger; they will still be susceptible to picking up worms, but will be more able to withstand a light challenge. There is also the benefit if turnout is delayed until mid-summer, by which time most overwintered larvae will have died off.

Strategic dosing with a wormer (anthelmintic): There are only three basic groups of broad-spectrum wormer used worldwide, to which the vast array of wormers available on the market belongs. These are:

Group 1: Benzimidazole (white drenches);
Group 2: Levamisole/Morantel (yellow drenches);

Group 3: Macrocyclic lactones, e.g. avermectins and milbemycins (clear drenches).

There is currently worldwide concern about the emergence of wormer resistance/tolerance, with many parasites showing resistance to one or more of the groups outlined above. This has been a gradual process, but there are instances now where resistance to all three wormer groups has been reported – and there are no new wormers on the horizon.

When wormer resistance to a particular group develops, it is irreversible, and applies to every member of the group, no matter what the formulation or manufacturer. As a result, there are now recommendations in many countries around the world to avoid further resistance developing, and to preserve those wormer groups that are still effective. There are some broad principles that are worth stating, although the inexperienced reader is advised to seek the advice of their veterinary surgeon on how best to proceed:

- *Administer the correct dose:* If you are able to weigh your goats to gain an accurate weight, this will ensure that you do not underdose, assuming your dose calculation is correct. If you are dosing a number of goats, dose to the heaviest goat, not the average weight – dosing to the average weight means that you underdose the heaviest. Ensure that your dosing equipment (particularly if it is an automatic drenching gun) actually delivers what the calibration states. By using a product licensed for use in goats, you can follow the manufacturer's recommendations regarding the actual dose regime to follow. If you are using sheep products, you may have to increase the dose rate of Groups 1 and 2 by a factor of between 1.5 and 2.0.

- *Use the minimum number of treatments:* Increasing the frequency of treating with anthelmintics increases selection pressure, and will lead to increased prevalence of resistant worms. Each unit should establish a control programme based on the minimum of doses. Avoid dosing routinely – for example, at 3–4 week intervals – unless the goats actually need worming, and have developed typical clinical signs. By delaying a wormer dose until it is necessary (ensuring that goat welfare is not compromised), you will delay the development of resistance in the worms on your unit. Large amounts of money are wasted each year on wormers that are simply not needed, and unnecessary use will also increase the development of wormer resistance.

- *Rotate the anthelmintics group annually:* This is a useful approach. Using the same group year on year will increase the selection pressure for the development of resistance.

- *Avoid introducing resistant worms:* Remember that purchased goats or sheep,

or sheep on tack or loan, could potentially bring resistant worms on to your holding. Many sheep flocks in the UK are currently advocating the use of a quarantine drench, involving the simultaneous administration of two wormer groups, together with a period of quarantine to allow resistant worms voided to be passed away from the pasture itself – ask your vet about this procedure.

- *Develop a farm- or unit-specific worm control programme* in consultation with your veterinary surgeon. No two units are identical, and it follows that no two worm control programmes should be, either! Consider the use of worm egg counts (*see* earlier) to give some indication of when a dose may need to be administered.

Haemonchus contortus (Haemonchosis)

Although this worm has already been described as a parasite of the goat's (and sheep's) abomasum, its life cycle and pathological effects are significantly different to the other stomach and gut worms to give it special mention.

It is a large roundworm, commonly called the 'Barber's Pole worm' due to its shape; the adults measure up to 3cm (1in) in length. It is essentially a blood-sucking worm, and if many thousands are present in the abomasum of an infected goat, they are capable of causing up to 200ml of blood loss daily, and hence a severe anaemia.

Life cycle: The life cycle is similar to that already described, but in climates such as the UK, the cycle is usually an annual one. Worm eggs are passed out in the spring, and develop to infective larvae in a few weeks. Rapid development of high numbers of larvae can occur during very warm weather, and severe disease with anaemia can result in very sick goats. Later in the season, many of the larvae picked up become dormant in the wall of the abomasum, bursting out again in the spring to continue their development (and produce further disease).

Resistance to Wormers

Remember it is the worms that have developed a resistance to the wormer, not the goats. You can buy in resistant worms in replacement goats, or your goats can move on to pasture grazed by other goats or sheep, which is heavily contaminated with a resistant worm population.

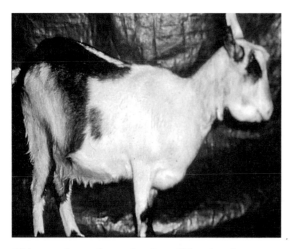

This goat has a heavy burden of Haemonchus; *note the swelling under its jaw (known as 'bottle jaw').*

Tapeworms in the gut, although large, are mainly harmless.

Clinical signs: Affected goats become weak and breathless when moved, and look pale around the mouth and eyes; severely affected goats may develop a fluid swelling under the throat (so-called 'bottle jaw'), which is related to anaemia and loss of protein. Scouring is not a feature of this worm, and disease caused by *Haemonchus* may be overlooked, as the signs are not typical.

Treatment: The worm is susceptible to most wormers available, but resistance to the white drenches in particular is a problem.

Tapeworms

Moniezia: Tapeworm segments may occasionally be seen in faeces, and are shed as part of the life cycle of the cestode (tapeworm) *Moniezia* spp. They can be quite large, and may wriggle around in the faeces, and can be quite alarming. In fact they look far worse than they are, and rarely cause any illness in goats, but live in 'perfect harmony' with the gut and its content. Rarely, very large numbers may cause problems in young kids, leading to poor growth rates, a pot-bellied appearance and constipation. Complete blockage to the gut will lead to signs of colic.

Metacestodes: There is a group of cestode tapeworms that is found in the gut of carnivores, particularly dogs, cats and foxes. Their

life cycle is referred to as a 'two-host' life cycle, because tapeworm segments are passed in the faeces of the carnivore, the segment then disintegrates releasing eggs, that may then be picked up by grazing herbivores (sheep mainly, but occasionally goats). The eggs hatch, liberating the parasite into the gut of the herbivore. These cysts then migrate to various locations in the body via the bloodstream. Many cause no more problems and are well tolerated by the goat, but may be seen at post mortem.

Very occasionally liver damage can occur if large numbers of cysts pass through. *Cysticercus tenuicollis* passes through the liver, and cysts may be found in the peritoneal cavity, but are for the most part innocuous. *Hydatid* cysts are occasionally seen in goats, but tend to be geographically confined to certain areas; the condition was a big problem in sheep in Wales, but has now largely been controlled. *Hydatid* cysts in sheep and goats rarely cause illness, even with heavy burdens. The main concern historically with hydatid disease has been the risk to humans, as cysts can develop in a number of body organs. The risk to humans is not from a goat, however, but from the original source of the tapeworm cyst – namely the carnivore, and principally the dog. Carnivores become re-infected by eating sheep or goat viscera containing the cyst, thus continuing the life cycle. Control is based on a regular and effective worming programme (for tapeworms) in dogs, and ensuring that sheep or goat carcases are not left out to be eaten by dogs or foxes.

Coccidiosis

A protozoal (single cell) parasite, more specifically referred to as *Eimeria*. There are many species of coccidia that affect a wide range of species; it has traditionally been a problem in the poultry industry, for example. It is important at the outset, however, to remind the reader that all *Eimeria* spp. are host-specific (with very few exceptions), and therefore the coccidia that affect chickens will not affect goats, and vice versa, and this rule can be applied to all species.

All goats can carry coccidia in their gut, and pass them out in their faeces (the sign of a well-adapted parasite). Disease is only seen in younger goats, however, as immunity will develop after exposure, sufficient to prevent clinical disease in older animals, but not sufficient to prevent the parasite multiplying in the gut, thus acting as a source of infection for another goat and perpetuating the cycle.

Life cycle: The life cycle is purely goat-to-goat, via faecal material. Adult goats with low levels of parasite in their gut perpetuate the parasite life cycle from year to year, although the free-living parasitic form has a thick cyst wall (the oocysts), and can remain in the environment if pens are not properly cleaned out. Susceptible kids will pick up infection from their environment; the coccidia invade the gut wall, and go through a massive multiplication phase to the extent that one oocyst ingested can result in 1,000,000 oocysts being excreted, thus causing heavy contamination. Problems often develop when successive kids born go through the same building, with minimal clean bedding being provided; thus earlier kids pick up low levels, and although they themselves remain fit, they nevertheless rapidly contaminate their environment. They can become re-infected themselves, but their immunity will partially protect them – but later kids are

This goat has suffered from severe coccidiosis. Note the faeces around its tail and back.

exposed to high levels, and very severe disease can result. This life cycle can be completed in housed or grazing goats.

Clinical signs: Diarrhoea, often with flecks of blood and mucus; many kids strain constantly. Affected kids rapidly lose condition, have a staring coat, and a dirty back end. In severe infection kids may be found dead. Although clinical signs are fairly typical, disease can be confirmed by the demonstration of high levels of coccidia in a faeces sample, or by post-mortem examination.

Treatment: There are few licensed products to treat this condition, but discuss treatment strategies with your veterinary surgeon.

Prevention and control: There are a number of 'rules' regarding rearing kids in order to avoid coccidiosis; these include:

- Clean pens out on a regular basis, trying to build a 'break' if rearing kids on a regular basis, to avoid any build-up of infection. Use a disinfectant that has a declared 'oocidal' effect – that is, it kills coccidial oocysts.
- Avoid overcrowding kids.
- Avoid having young goats of different ages in the same pen or paddock.
- Ensure that you have food and water bowls, preferably raised off the ground, to prevent goats defecating into them – provide covers if necessary.

- If there is a confirmed problem, discuss the possibility of routine medication with your vet.

Cryptosporidiosis

Cause: A small protozoal parasite closely related to coccidia. One important difference, however, is that, unlike coccidia, cryptosporidia are not host-specific, with most infection in farm animals caused by the same species (*C. parvum*).

Age group affected: This is mainly seen in young kids (referred to in Chapter 6 as a cause of *neonatal diarrhoea*), between one and four weeks of age, by which time most kids will have developed an immunity. Disease can be particularly severe in kids that have received insufficient colostrum. Older goats can, however, act as a source of infection for younger kids (as carriers).

Life cycle: As already stated, the parasite lacks host specificity and, as a result, kids can be infected from a wide range of sources, including other goats (and infection can be introduced into a clean herd by the purchase of infected carrier animals), calves or lambs. Mice and rats may also carry infection, and the oocysts are very resistant and can survive in contaminated pens or paddocks.

Once ingested, the oocysts invade the gut wall, go through several multiplication phases, resulting in the passage of high numbers of oocysts in faeces, thus acting as a further source of infection.

Signs: Common signs include watery diarrhoea, dehydration and wasting. Death can be a sequel to heavy infestation as a result of gut damage and dehydration, resulting in renal failure.

Laboratory tests can demonstrate the typical oocyst in fresh faeces, or at post-mortem examination.

Treatment: There is a specific product (halfuginol) that has been used with a moderate degree of success in calves, but treatment is mainly symptomatic – that is, it merely controls the diarrhoea and dehydration. Discuss this with your vet.

Prevention: Is based mainly on ensuring that all kids receive adequate colostrum. There is merit in removing kids from the dam's environment shortly after birth, and rearing them artificially if there is a particular problem with cryptosporidia (the dam acts as a source of infection). Keeping group size to a minimum, with regular cleaning of buildings between batches, is also important. Isolate any kid with diarrhoea.

Public health: Cryptosporidia can infect humans who are in contact with kids (and lambs or calves) or their faeces. Those particularly at risk include young children, the elderly, and those who are on certain medication that may suppress their immune system (so-called immunosuppressive drugs). Care should be taken in any situation where young children are exposed to goat kids, such as visiting schoolchildren, or in open farms or parks where visitors may have contact (Chapter 17).

OTHER CONDITIONS AFFECTING THE GUT

Clostridial Enterotoxaemia

There is a group of bacteria referred to as *Clostridia* spp. that are part of the normal gut flora, aiding in the process of digestion. They are also capable of forming spores that can survive for long periods in the environment. A number of *Clostridia* are capable of causing disease in goats, normally when the balance between the host immunity or the weight of infection is upset. In sheep there are approximately eight different clostridial diseases that most shepherds vaccinate against (using the so-called '4-, 7- or 8-in-one' vaccines, named after the number of clostridial diseases that are protected by the vaccine).

Goats suffer from very few of the diseases caused by *Clostridia*, although they are particularly susceptible to disease caused by *C. perfringens* Type D. Tetanus (Chapter 10) may also be a problem, although the gas-gangrenous conditions such as blackleg, malignant oedema, black disease and braxy are rare.

Clostridium perfringens Type D Enterotoxaemia

Clostridium perfringens Type D is the organism responsible for causing pulpy kidney in sheep, although the presentation in goats is quite different. It is important to stress at the outset that this infection is due almost entirely to an imbalance between the immunity of the goat and the level of infection of what is a normal inhabitant, in its gut. Precipitating factors include a sudden change in diet, overfeeding cereals, introduction of lush grass or other vegetation and stress. The author has encountered the problem on large units where shy goats from small units appeared to develop the condition purely as a result of the stress of integrating into an unfamiliar environment.

When the organism multiplies rapidly in the gut, it produces increasing amounts of a powerful toxin that is absorbed into the bloodstream causing toxaemia, with damage to brain, heart and blood vessels, in addition to severe damage to the gut wall itself.

Age group affected: This disease can in theory develop at any age, but is less common in young kids.

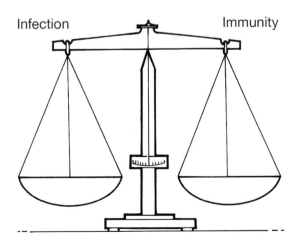

Clostridial infection – a delicate balance!

Clinical signs: There are three main presentations:

Per-acute: Sudden death is not uncommon, and goats can be found dead without any other apparent signs.

Acute: The main presenting sign with this presentation is diarrhoea and dysentery (blood in the faeces), often with a large amount of mucus and shredded gut wall passed. As the condition progresses, the rapidly increasing toxaemia results in other signs including collapse and shock, and nervous signs including convulsions; affected goats may be very vocal, suggesting severe abdominal pain.

Chronic: Reported in larger units, toxin produced in the gut damages the gut villi, leading to the passage of pasty faeces, and wasting as a result of damage to the normal absorption processes that occur in the gut (malabsorption).

Disease in live animals is fairly characteristic, and is normally confirmed on clinical appearance and the elimination of other possible causes of dysentery and diarrhoea. Faeces samples sent to a laboratory may be tested for the presence of toxin. The lesions seen at post mortem are fairly characteristic, and toxin can be demonstrated in gut content.

Treatment: Success depends on the stage the disease has reached before treatment begins. Once the toxaemia becomes overwhelming, treatment is unlikely to be effective. Your vet must therefore be contacted as a matter of urgency to ensure that a rapid treatment regime is begun. Fluids orally or intravenously are always effective, with pain relief and anti-inflammatory agents. Pulpy kidney antiserum (containing *Clostridium perfringens* toxin) may also be useful.

Prevention: There is an effective vaccine available. As already stated, goats suffer from only a small number of the clostridial diseases that affect sheep, and your veterinary surgeon will normally advise that you use a '4-in–1' vaccine. Goats do not appear to be able to produce such good immunity to clostridial vaccines as sheep, another reason

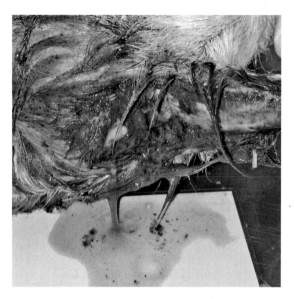

Blood and mucous in the diarrhoea of a goat with acute clostridial infection.

for using a vaccine with only minimal components, thus ensuring that the goat's immune system is directed to the important clostridia, and in particular *C. perfringens* Type D. Manufacturer's recommendations must be followed, and this will include two initial doses, with a normal recommendation to boost every six months.

Ensure that you introduce new dietary components gradually, and that you recog-

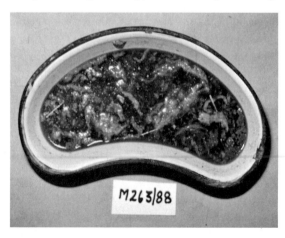

Gut content – almost pure blood – from a goat dying from clostridial enterotoxaemia.

nize 'stressed goats', and take steps to reduce the stress (coupled with vaccination to ensure that the goat is protected).

Johne's Disease (Paratuberculosis/ 'MAP')

Cause: A mycobacterial organism referred to as *Mycobacterium johnei* or, in recent years, as *Mycobacterium avium* subsp. *paratuberculosis* (now commonly referred to as MAP). Infection with this organism causes a chronic inflammatory bowel disease referred to as Johne's disease. The same organism can cause similar pathology in the gut of both cattle and sheep, and care must be taken to ensure cross-infection on a mixed stock farm, if disease is confirmed in one species.

Age group affected: It is a chronic disease, picked up by young kids normally in the first six months of life, after which they appear to become resistant to picking up new infection. The disease then appears to lie dormant until gut damage develops slowly and insidiously, with clinical signs normally apparent from two years of age and older.

Clinical signs: Early subtle changes may go unnoticed; these include a gradual deterioration in milk yield and quality. As the condition begins to develop, however, goats will visibly waste, appear anaemic (with pale, visible mucous membranes) and lethargic, and their milk may well dry up. Unlike disease in cattle, however, in which 'pipe stem diarrhoea' is a classic feature, diarrhoea is not a feature in goats until the terminal stages, and faeces will normally remain pelleted. Oedema swelling under the jaw – known as 'bottle jaw' – may be a terminal feature, related to very low body protein reserves.

Confirmation of disease: A number of conditions can cause dramatic weight loss in goats, and some have already been mentioned in this chapter. Your veterinary surgeon should be consulted, and will thoroughly examine your goat to rule out any obvious underlying clinical problems. Laboratory tests are required to confirm Johne's disease however, but these are by no means simple tests, and

By gently pushing down on the lower lid, the conjunctiva can be examined; it should be a healthy pink in colour – this goat is very anaemic.

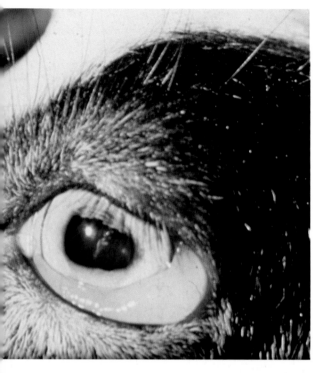

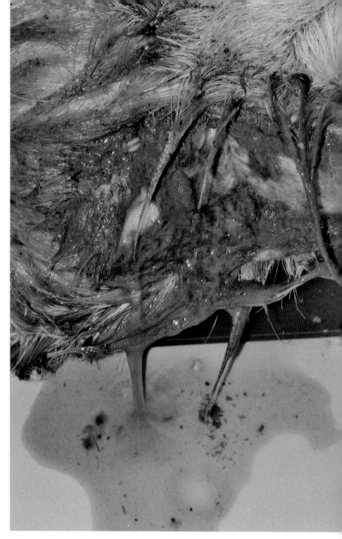

Severe diarrhoea with blood and mucous around the anus of a goat severely affected with clostridial infection.

This sign on the fence of a high health-status pig unit shows just how seriously the pig (and poultry) industry takes disease control/biosecurity.

This Pygmy goat is showing hair loss and abscess formation; a number of causes need to be considered.

Ultrasound scan – this procedure can confirm pregnancy and the number of foetuses carried.

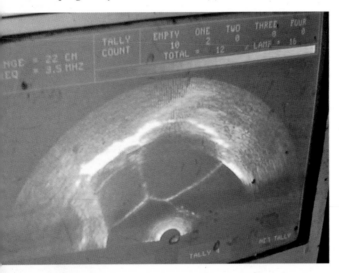

Aborted kids from the same doe; note the disparity in size (and the absence of the placenta!).

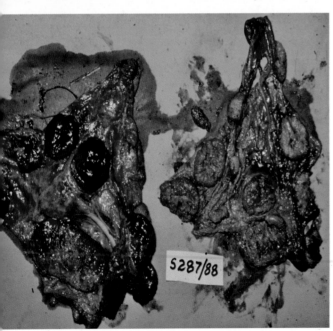

This placenta is from a goat with confirmed Chlamydial abortion; note the thickening of the area between the cotyledons (buttons).

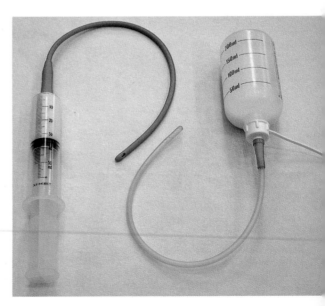

Two adaptations of a stomach tube (note its blunt-ended shape).

Infra-red lamps are a useful way of warming a kid, but take care not to cause the kid to overheat!

Some goats may bleed profusely after the horn is removed; this wound is being cauterized with a hot iron.

This kid has been recently disbudded.

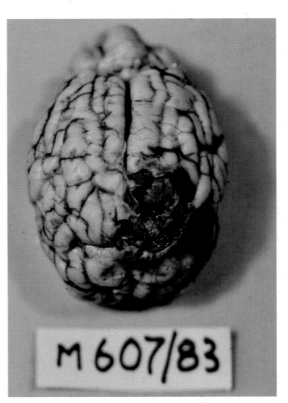

This goat kid was disbudded in unskilled hands. An iron was used that was too cool, so it was held against the head for too long a period of time, with the result that heat was transferred across the skull, literally 'cooking' a small area of brain.

This goat died from acute acidosis; note the large amount of grain in its rumen. The content will be very acid.

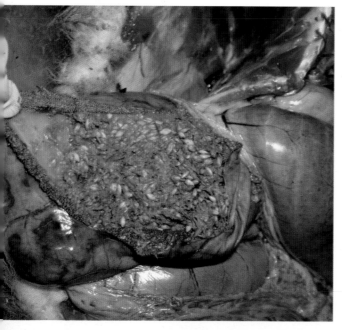

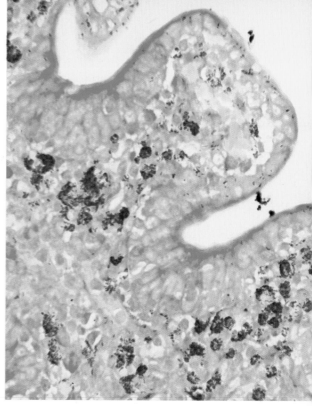

Special staining has shown the typical acid-fast organisms in the gut wall (under a microscope).

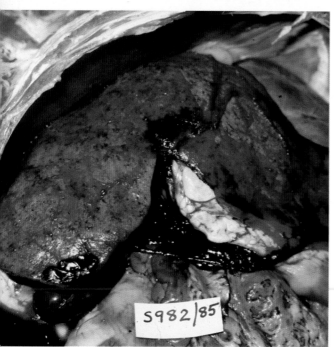

The liver of a sheep dying from acute fluke (a goat liver would be similar) – note the haemorrhage that has occurred as the immature fluke burrow through the capsule.

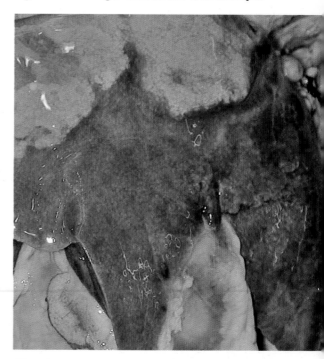

These lungs were severely damaged by pneumonia; note the 'normal' pink tissue and dark, consolidated lung.

This brain has been removed and incised, allowing the gid cyst to escape under pressure.

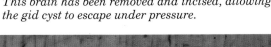

The protective wall covering the sensitive laminae has been lost as a result of severe footrot; this is referred to as 'thimbling'.

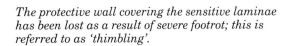

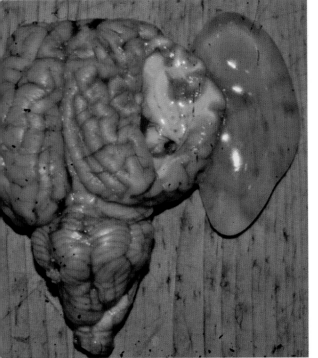

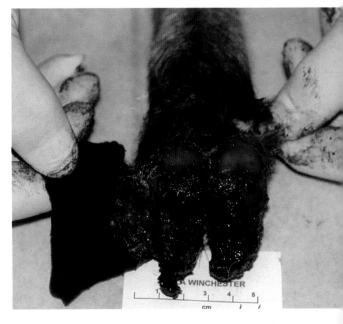

This goat has listeriosis; note the drooping ears and 'dull' expression.

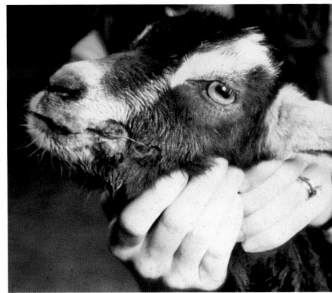

This kid shows a typical orf (parapox) lesion on its lips.

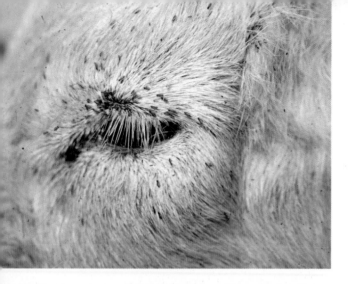

This Saanen kid has a heavy lice burden clearly visible around its eyes.

A typical 'pink eye' appearance with severe conjunctivitis.

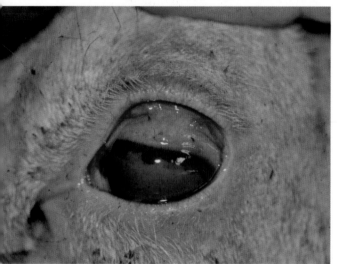

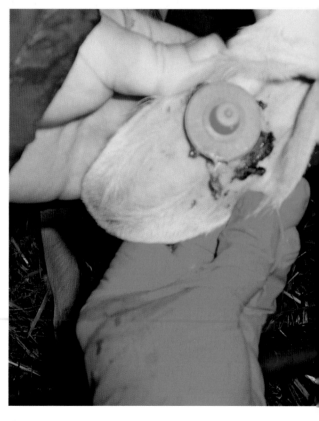

Fluorescent dye has been placed in the eye – it will stain damaged cornea, particularly if there is an ulcer.

Goats appear to be more susceptible to local reactions around eartags than sheep or cattle.

This udder is affected with acute gangrenous mastitis; note the blackened appearance and oozing of blood-tinged fluid from its surface.

These small pimples on the udder are caused by Staphylococcus aureus, and can predispose the goat to develop mastitis.

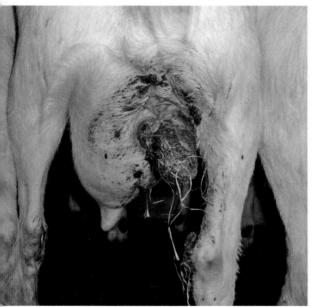

Part of this udder has literally 'sloughed away', leaving a large raw area. This goat should be under veterinary care.

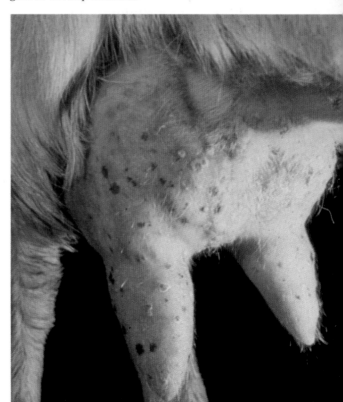

These slug pellets were seen lying around a bonfire built to dispose of the bags.

Pieris – an attractive member of the rhododendron family, but highly toxic to goats.

This old can of putty (very high in lead) was found after a tree was pulled up. Note the teeth marks!

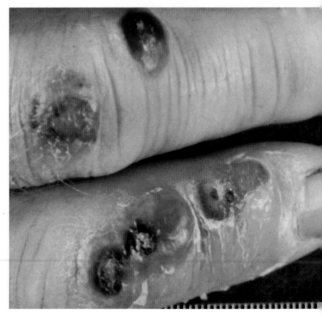

Orf lesions on the fingers of a goat owner whose goat had developed orf.

must be applied and interpreted with care. Examination of a faeces sample may demonstrate the typical appearance of mycobacterial organisms microscopically, mainly by their so-called acid-fast staining characteristic. A blood sample can demonstrate antibody in the later stages of infection. Post-mortem examination of a typical case will show characteristic thickening of the gut wall, with an associated enlargement of the lymph nodes.

Can it be treated? There is *no treatment*, and once the condition is recognized (by which time most goats are already beginning to deteriorate), the goat should be culled on humane grounds before the condition worsens. Bear in mind that offspring of affected goats may also be infected, and this is a further reason for keeping good records.

Control and prevention: Depending on the type of unit you have, there are a number of options open. If the unit is commercial, ask your veterinary surgeon to put together a control programme specific to your own unit. This should include:

- Culling clinical cases (+/- post-mortem examination, depending on the stage of the control programme).
- Use of laboratory tests, such as examination of faeces and blood, to identify infected goats before they become clinical, and shed infection at high levels.
- Identify offspring of infected goats, and cull (many will be infected from close contact with their dams while suckling).
- Consider use of a Johne's disease vaccine – administered to young neonatal kids, it has been used very effectively in the UK.
- Consider snatching kids from the adult goat environment, thus reducing the chance that infection may be picked up.
- Avoid the use of pooled colostrum on infected farms – one infected adult goat may infect a colostrum pool, thus infecting a whole group of kids.
- If the goat unit is clear of infection, then ensure that any purchased goats come from herds known to be free of infection wherever possible. The available laboratory tests are not really robust enough to be used as screening tests in apparently healthy goats, but discuss this with your vet.

Decisions are more difficult with pet goats and Johne's disease. As already stated, there is no treatment, but your vet may be able to prolong the goat's life with supportive therapy, including vitamins and minerals. Nevertheless, the welfare of the goat must be paramount, and once its quality of life deteriorates, then euthanasia must be the only option available.

Risk to humans: There is still some debate as to a possible link between Johne's disease in cattle (and goats: it is the same organism) and Crohnes disease in humans.

Salmonellosis

As already described (*see* Chapter 6), Salmonellosis is not particularly common in goats, but it may occur most often in animals under stress (particularly in the post-kidding period). Infection is normally picked up from an outside source such as rats, mice, wild birds, leaking septic tanks, wheels of vehicles or wellington boots contaminated with, for example, cow faeces, in which the condition is more prevalent.

This disease can be particularly severe, with dysentery, fever, dehydration and shock developing rapidly. Consult your vet as a matter of urgency; the disease is normally confirmed by laboratory examination of faeces or a rectal swab. Treatment must be rapid and decisive if it is to be successful, and is based on the administration of a suitable antibiotic and fluids.

Because the environment can become heavily contaminated, and be a danger to other goats, a full cleaning and disinfection programme is important, and affected goats must be placed in quarantine until the diarrhoea has cleared up.

Remember the organism is referred to as a 'zoonotic organism': this means it will readily transfer to humans in direct contact, or can be spread indirectly by the consumption of raw milk.

DISEASES OF THE LIVER

Liver Fluke (Fasciolosis)

A flatworm trematode parasite known as *Fasciola hepatica*, or more commonly referred to as 'liver fluke'. The liver fluke has a complex life cycle, as shown in the diagram, and is a parasite shared by most grazing ruminants including both sheep and cattle, but confined mainly to wetter areas.

Life cycle: The adult liver fluke lives in the gall bladder and bile ducts, where each fluke can survive for up to four years if the goat is not treated. The fluke lays eggs in the bile ducts and gall bladder, and these are then carried via the bile and bile duct into the gut, where they pass through, eventually being deposited on the pasture. For the life cycle to continue, however, the eggs must hatch and the next developmental stage (the miracidium) must then penetrate the body of a very specific mud snail, *Lymnaea truncatula* (hence the

reference to wetter areas of the country). The developing parasites then go through further developmental and multiplication stages in the snail, before re-emerging as free-living forms that rapidly encyst, and become available on the grass to be eaten by grazing goats, sheep or cattle.

Once inside the gut, the immature fluke pass out of the wall of the gut into the abdominal cavity, and make their way to the liver surface where they literally 'bore' their way through the capsule into the liver tissue, causing considerable damage with large burdens. Once back in the liver, they find their way to the bile ducts and gall bladder to begin the life cycle again.

Clinical signs: Signs of disease are referred to as either acute or chronic, and present as two fairly clearly defined syndromes:

- *Acute disease:* Caused by large numbers of immature fluke boring into the liver surface, often causing haemorrhage. Signs include anaemia, abdominal pain

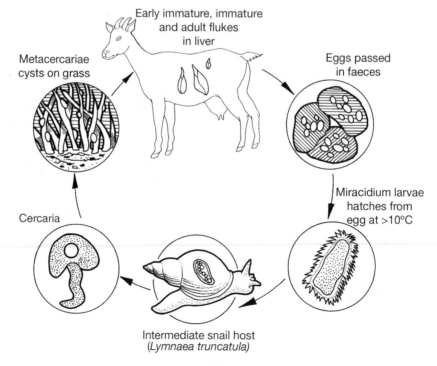

Liver fluke life cycle.

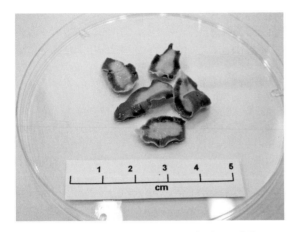

Liver fluke parasites; they live in the liver, bile ducts and gall bladder.

and weakness. In severe cases affected goats may be found dead. Disease is most often encountered in late summer/early autumn.

• *Chronic:* Related to the burden of mature fluke found in the bile ducts and gall bladder, causing chronic long-term liver damage. Signs include weight loss, anaemia, and bottle jaw (fluid accumulation under the skin linked to low body protein levels). This disease is encountered at any time of year, but particularly in the autumn and spring.

Treatment: Before embarking on an expensive treatment programme, ensure that fluke is the underlying problem. Disease will mimic a number of other conditions including worms, *Haemonchus* and Johne's disease. Any fluke egg found in a faeces sample indicates a mature burden, but it takes 8–12 weeks for eggs to appear in faeces after infection is picked up, so this test is only reliable in the later stages of infection. Blood samples may confirm liver damage, and low body protein levels. Post-mortem examination will rapidly confirm a diagnosis. There are a number of products available, so consult your vet as to the most suitable treatment regime.

Prevention: From looking at the life cycle, it is clear that there are strategic points where control measures can be put in place. By avoiding having goats in areas that are dry for at least part of the year will remove the snail habitats, and the snail is necessary for the life cycle to be complete. Fencing off wet areas such as ponds, open ditches, riverbanks, marshland or even leaking water troughs will all help. The second alternative, if there is no chance of avoiding the snail habitat, is to dose the goat with an effective flukicide at strategic times of the year, to coincide with periods of greatest risk. A fluke forecast is available in many countries, giving information to livestock-keepers on the timing and potential severity of fluke risk.

Liver abscesses in kids: *See* Chapter 6 (navel and joint ill).

Tapeworm cysts: *Cysticercus tenuicollis* and hydatid cysts can both migrate through the liver (*see* earlier in this chapter under 'Tapeworms').

Tumours: The liver is a fairly common site for malignant tumours to spread from their primary site. Signs are usually fairly vague, and their presence is normally confirmed at post-mortem examination.

Liver Fluke Disaster

The worst single incident of liver fluke encountered in goats by the author occurred in a group of six goats that were tethered around a duck pond, when grass was in short supply during a particularly dry summer. There was plentiful grass present, but sadly also a heavy population of mud snails, clearly visible in the mud. A total of three out of the six goats died over a weekend with acute fluke, and post-mortem examination showed a severe internal haemorrhage from the damaged liver capsule. The remaining animals survived, but despite treatment, one other goat developed massive liver damage, and died later. Sheep had been grazing in the same area previously, and they were thought to have infected the snails. The goats were denied their normal browsing activity, and forced to feed in a confined area, wherein a heavy snail population containing a high level of immature fluke was also present.

CHAPTER EIGHT

The Respiratory System

In this chapter we will consider those problems affecting nostrils, nasal chambers, sinuses, larynx, trachea and lungs.

The clinical disease that is seen may vary from a mild nasal discharge to sudden death from an overwhelming pneumonia, and it is possible to see the full range of clinical signs if an infectious problem develops in a group of goats. Certain problems develop rapidly and run their course, others can be chronic and longstanding. The most common clinical signs to look out for include nasal discharge, coughing, sneezing, dyspnoea ('difficult, often noisy breathing') and hyperpnoea (rapid breathing) with pyrexia and inappetance. Many of these are obvious, but take time to observe your goat's breathing if you are in doubt; it is often useful to compare their breathing with others. As the breathing becomes more difficult, the goat will gradually extend its neck out straight to aid the movement of air into and out of its lungs.

Difficulty breathing can also be seen in a range of conditions unrelated to the respiratory system, such as bloat, choke and acidosis (Chapter 7), hypocalcaemia (Chapter 5), anaemia (Chapter 9) and some poisonings (Chapter 16). It may also be a feature of heat stroke (occasionally reported in show goats in tents on a hot day). Halitosis (bad breath) is normally related to digestive problems such as tooth and gum disease.

INFECTIOUS CAUSES OF RESPIRATORY DISEASE

It is important to emphasize that many infectious causes of respiratory disease may well be multi-factorial – that is, there is more than one underlying cause. It may be, for example, that one or more infectious agents are involved, but environmental factors such as ventilation, dust levels and so on, all play a part.

Pasteurellosis

The main bacteria involved are *Mannheimia haemolytica* (previously referred to as *Pasteurella haemolytica*) and *Pasteurella multocida*. There are a number of differing strains of *Mannheimia* described, known as 'A' and 'T' strains or serotypes. In the UK, most disease is caused by A1, A2 and A6, with T strains rarely involved, being more often associated with disease in sheep. The 'A' serotypes are common to disease encountered in both species, hence the possible spread between species if housed near each other. Most available vaccines give protection against a wide range of different serotypes.

It is important to emphasize from the outset that *Mannheimia* organisms can be found in the throat and tonsils of a high proportion of apparently healthy goats. For true disease to develop, the infection needs to move from its harmless existence in the upper airways to become established in lung tissue, where varying degrees of damage can occur. The trigger for this to happen is still poorly understood in both goats and sheep, but is likely to be similar in each. Disease will often follow a stressful insult, such as weaning, overcrowding, poorly ventilated or dusty buildings, or concurrent illness or injury. It can also follow specific damage to the upper airways that some other infectious respiratory agents can cause, for instance *Mycoplasma*, *see* below.

Is it infectious? Although the organisms will undoubtedly spread between goats in an enclosed space, most cases appear to develop as a result of stress or concurrent disease, in animals already carrying the organism.

Age group affected: Pasteurellosis can develop at any age, but appears to be a particular problem in youngsters. That said, the author was involved in an outbreak where thirty deaths occurred in adult goats, admittedly under considerable stress having recently been transported a long distance, and mixed with goats from at least five other sources, carrying a whole range of differing micro-organisms, all trying to adapt to a new management and feeding system!

Disease is more likely to be seen in housed goats than in those kept outside.

Clinical signs: In the early stages, the goat will run a slight temperature and go off its feed. As lung damage develops, the breathing will become more laboured and rapid, and the head will be held out straight. There may be a cough and nasal discharge, with the nostrils crusting over. Cases need to be caught early, so contact your veterinary surgeon if you are concerned; he/she can assess the degree of lung damage by listening to the chest with a stethoscope. The post-mortem picture is quite characteristic.

Treatment: Your vet will normally prescribe antibiotic medication, and may also suggest that you treat other in-contact animals as a precaution. Keep affected goats in an airy building or pen with shelter from wind and rain. It is important that they can be caught easily to administer medication, since the stress of having to chase them around will only cause them further distress.

Prevention: Although there are vaccines available to prevent pasteurellosis, discuss their use with your veterinary surgeon. There is probably no need to vaccinate goats against pasteurellosis routinely, unless either there has been a previous history of infection on the unit, or emergency vaccination can be part of a control programme in a new outbreak that becomes protracted. Although some manufacturers market a combined clostridia/pasteurella vaccine, the author would strongly recommend that separate vaccines are used for each organism, to enable the goat to respond to its fullest potential – that is, to mount a good immune response. The vaccine normally requires two doses four to six weeks apart, followed by annual booster doses if deemed necessary.

In pregnant does this booster is best given during late pregnancy to boost colostral levels of antibody. Kids that receive sufficient colostrum from fully vaccinated and boosted dams will be protected for approximately one month, but will need boosting if disease is a problem in young kids. It may be worth vaccinating show goats against pasteurellosis even if disease has never been encountered previously (*see* below).

Other management procedures that are important in preventing a possible problem with pasteurellosis should include:

- Avoid overcrowding, particularly in housed goats; ensure that they have a sufficient lying area and trough space.
- Avoid housing goats in poorly ventilated buildings: air should flow in freely from the sides of the building above the goats (thus avoiding draughts), and then be allowed to escape at a high point in the building to keep the air fresh. Warm, exhaled air will rise, and should then be removed if air is to circulate correctly.
- Avoid mouldy/dusty feed and bedding – mould spores or dust can irritate the upper

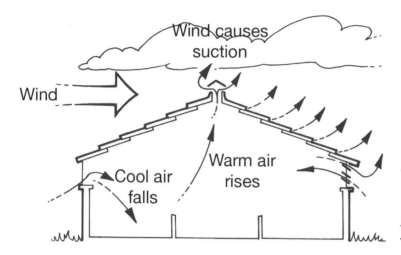

A simplified diagram of natural ventilation to show the basic principles of ventilating a building, to prevent respiratory disease, yet avoid draughts.

airways, potentially allowing pasteurella organisms to multiply.

- Assess the possible role of other infectious agents, particularly when mixing goats together. If one group of goats is coughing, and the other is not, keep them separate – the mixing of different micro-organisms can result in severe disease. Quarantine incoming animals before mixing them with your own goats; this is a particular problem at shows as goats are mixed together.
- Heat stress can predispose to infection; this may be a problem in show goats already stressed from being transported to and from a show, but hot weather in a poorly ventilated tent may make things worse.

Mycoplasmosis

Mycoplasma organisms are biologically mid-way between a bacteria and a virus. There is a range of mycoplasma organisms that are capable of causing illness in goats, the most serious disease in goats worldwide being contagious caprine pleuropneumonia (CCPP), confined mainly to parts of Africa and Asia.

In the UK, respiratory disease caused by mycoplasma is generally mild, but as already stated, can be a precursor to more severe respiratory disease in combination with pasteurella. The main organisms involved are *Mycoplasma ovipneumoniae* and *Mycoplasma arginini*. Clinical disease is usually confined

to young kids, in which outbreaks of coughing and nasal discharge can occur. Disease is often self-limiting but, if in doubt, consult your veterinary surgeon.

'Show cough': Many goats returning from weekend shows appear to develop a mild cough, and may be off their food for twenty-four to forty-eight hours before recovering. There is some thought that this could be related to mycoplasma infection.

Caseous lymphadenitis (CLA): *See* Chapter 9.

Viruses

There is little evidence to suggest that any of the viruses linked to respiratory disease in cattle worldwide are able to cause natural disease in goats, although some have been transmitted experimentally.

Caprine arthritis encephalitis (CAE): This virus can cause chronic insidious lung damage, referred to as a pneumonitis. CAE is covered in detail in Chapter 11.

Lungworm

There are two types of lungworm causing illness in goats: *Dictyocaulus filaria* and *Muellerius capillaris*. Both are shared parasites with sheep.

Is it infectious? Each parasite has a slightly different life cycle. That of *Dictyocaulus* is direct, with the worms found in the trachea and airways; immature larvae are coughed up, swallowed, then passed out in the faeces. Following two developmental stages at grass, the larvae are picked up again, to complete the life cycle, finding their way back to the airways. The life cycle of *Muellerius* is slightly more complex. The worm is microscopic, and lives in the alveoli, the small airsacs that are found at the end of the airways, millions of which make up the lung tissue. Larvae are again coughed up and swallowed, passing out with the faeces on to the pasture. They then need to pass through, for example, a slug or a snail, and go through further developmental stages before re-emerging to be eaten by a grazing goat; thus the life cycle continues.

Age group affected: As stated in Chapter 7, goats have a poor ability to produce an immunity to nematode parasites, and as such, disease can be seen at any age, although it is likely that it would be most severe in young goats.

Clinical signs: This disease is not very severe unless there is a heavy infection. The signs seen are fairly common to both types of lungworm-provoked illness, and include weight loss, a drop in milk yield, a mild cough, and in mild cases poor exercise tolerance (for example, when driven the goat is seen to have difficulty keeping up with the others), to severe dyspnoea in heavy infections. Disease can be confirmed by the laboratory examination of faeces, or at post-mortem examination.

Treatment: By the use of an anthelmintic; consult your vet to rule out other causes of respiratory disease.

MISCELLANEOUS CAUSES OF RESPIRATORY DISTRESS

Inhalation Pneumonia

This condition can occur when food material or liquid such as a drench enters the trachea instead of the oesophagus; the material can then be inhaled deep into the lung tissue, causing immediate distress. If

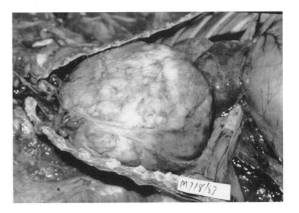

A large thymic tumour, or thymoma, found at post-mortem examination. Its size occupies much of the thoracic cavity.

large quantities of fluid are involved, death can quickly follow, with the goat literally 'drowning'. With smaller amounts, the goat may survive the initial insult, and appear to improve, only to become very sick over the next day or so, as bacterial multiplication occurs deep in the lung tissue, causing a severe broncho-pneumonia.

It is a common sequel to rhododendron poisoning (*see* Chapter 16).

If you suspect that inhalation of fluid has occurred, you can try swinging young kids upside down, which may help the fluid to flow back. This will be more difficult in older goats, but lifting the back legs higher than the front end may help, the same way that you would pick up a wheelbarrow to push it. Antibiotic cover may help to prevent secondary bacterial infection.

Drenching Gun Injury

In unskilled hands it is possible to cause damage to the back of the throat when using a drenching gun to dose a goat with, for example, a wormer. This may happen if a goat is not properly restrained and struggles during the procedure; it can also occur if the nozzle of the gun is bent out of its correct shape. The goat will initially show discomfort as if choking, but may then appear to recover. Secondary infection may then cause a local abscess to develop (often near the larynx), and the resulting pressure will again cause respiratory distress.

Tumours

Primary tumours of the lung are not very common. The lung is, however, a common site for secondary tumour deposits (metastatic tumours) to develop from primary tumours in a number of sites, normally in older 'pet' goats rather than commercial goats that tend to be culled at a younger age.

Thymomas – tumours involving the thymus (a lymphoid structure in the chest near the heart) – can be very large; the author has encountered one 'football'-sized thymoma occupying the majority of the thorax and causing progressive pressure on heart and lungs. The history was of increasing respiratory distress over a period of several weeks.

Winter Cough

This is a condition somewhat similar to hay fever in humans. It is seen mainly in older goats during the winter housing period, and is thought to be due to an allergic reaction to dusts, pollens and/or fungal spores in poor-quality, mouldy feed or bedding. Affected goats develop a chronic cough and an increased breathing rate, and they lose weight; they are often affected in successive years. When they are outside they breathe normally.

CHAPTER NINE

The Circulatory System

In the goat, the circulatory system rarely develops any specific abnormalities, with one important exception, although some conditions such as anaemia and jaundice are an indication of disease or illness in other organs.

The circulatory system includes the heart, the major blood vessels that take blood around the body, and the smaller vessels and capillaries that carry blood through all body tissue. It also includes the lymphatic system and associated lymph glands.

In broad terms the heart is merely a sophisticated muscular pump consisting of four main chambers (the atria or auricles, and the ventricles). The left side of the heart (pumping blood around the body) is separated from the right side of the heart (lung circulation) by a solid muscular wall. Sophisticated valves separate the atria and ventricles on each side. In a correctly functioning heart, blood passes through the lungs picking up oxygen, returning to the heart before passing into the general circulation, then back to the heart to repeat the process.

The blood itself consists predominantly of fluid or plasma, carrying a host of nutrients and body chemicals such as glucose, amino acids, proteins, enzymes, hormones, antibodies and electrolytes (sodium, potassium, calcium), together with the waste products of cell activity. Within this fluid is a suspension of cells, including:

- *Red cells* (*erythrocytes*), essential for carrying oxygen and carbon dioxide around the body via its iron-containing pigment haemoglobin.
- *White cells* (*leukocytes*) – there are a number of different types including neutrophils, lymphocytes and monocytes. These protect the body from infectious diseases such as bacteria and viruses, by attacking the micro-organisms in the bloodstream, or by involvement in the production of antibodies to fight disease.
- *Platelets* – these are the smallest blood cells, and are an important part of the blood-clotting mechanism.

Blood as a Diagnostic Aid

Your veterinary surgeon will often take blood samples to aid in arriving at a diagnosis, or to monitor disease. The table below gives the 'normal' or reference ranges for the main red and white cells in a healthy goat. Any deviation from this range may indicate, for example, anaemia, or infection. Levels of a whole range of biochemical parameters such as calcium, copper, ketones, albumin and urea are also important diagnostic aids, and damage to liver and muscle can result in the release of enzymes that can also be measured and compared with normal reference ranges. The blood also carries antibodies to disease that your goat may have been exposed to. Single samples are useful for the detection of antibodies to, for instance, CAE (*see* Chapter 11). Your veterinary surgeon may occasionally suggest taking two blood samples from a goat separated by a period of two weeks, checking for a rise in antibodies following an acute infection such as toxoplasmosis (so-called 'paired serology', using acute and convalescent samples). You may see your veterinary surgeon taking blood samples in containers with different colours or with different coloured stoppers to preserve certain components that may deteriorate as samples are transferred to a laboratory.

	Range	Units
Red cells	10 – 18	$\times 10^{12}/l$
Haemoglobin	8 – 15	g/dl (grams/ decilitre)
White cells	6 – 14	$\times 10^{12}/l$

Typical haematology reference ranges

The other important part of the goat's circulation is its lymphatic system. This system drains away fluid accumulating outside the blood vessels in the body tissues, via a series of lymphatic vessels and lymph glands or nodes, eventually draining back into the bloodstream near the heart. Lymph fluid is rich in the protective white cells, which multiply in response to infection either locally, such as following a cut or wound, or in a more generalized manner, for example as a result of a septicaemia. It is this increase in activity that results in the swelling of lymph glands that are often difficult to find. Those easily palpated include glands under the chin (sub-mandibular), around the base of the ear (parotid), and in front of the shoulder (prescapular). Lymph glands can also swell internally in response to infection, as the mesenteric glands that enlarge in Johne's disease (*see* Chapter 7).

DISEASES OF THE HEART AND BLOOD

Anaemia

The term 'anaemia' refers to an abnormality of the red cell series, with a reduced ability of the blood circulation to deliver vital oxygen to the body. The total number of red cells may be reduced in number, or there may be an abnormality of the haemoglobin content of the red cells. A goat may have to be quite anaemic before it begins to show signs of ill health (weakness, reluctance to move around and panting initially, followed by recumbency). Severely anaemic goats may be found dead due to cardiac failure. Anaemia can be detected in the live goat by looking at the mucous membranes of the eye in severe cases. Your vet may take blood samples to check.

The most common causes of anaemia in a goat relate to blood loss, although it can be due to a failure in replacement of red cells due to bone marrow abnormality, or to iron deficiency. Two conditions of the gut already described in Chapter 7 (Haemonchosis and Johne's disease) commonly cause anaemia if the condition becomes well established. Another parasitic cause of anaemia also referred to in Chapter 7 is acute liver fluke infestation, when many immature fluke burrow through the liver capsule, causing haemorrhage into the abdomen. A heavy burden of external parasites such as sucking lice (Chapter 12) can also result in anaemia. Mineral deficiencies such as copper deficiency (Chapter 10) and vitamin E deficiency (Chapter 12) are other possible causes.

Your veterinary surgeon will also consider blood loss due to trauma: for instance, the severing of a major artery; or there may be excessive blood loss from the vulva after kidding. Internal injuries, however, such as may result from a ruptured spleen or liver, may be less obvious.

A severely anaemic animal is a veterinary emergency: it will be shocked, and must be kept quiet and warm until help arrives. If there is an obvious source of blood loss such as a laceration, then apply gentle pressure with a clean cloth or handkerchief. Be aware that although a tourniquet is useful in preventing blood loss if applied above, say, a severed lower limb, it must be released periodically to allow blood to flow through the remainder of the limb, which could become damaged if deprived of a blood flow for too long.

Oedema

This condition is the build-up of fluid in the body that escapes from blood capillaries and lymphatic vessels. It can develop under the skin, for example as 'bottle jaw', when fluid accumulates beneath the jaw, or in dependent parts of the body, particularly the lower limbs. This swelling will 'pit', so that your finger literally leaves an indentation if you gently prod the affected area. Oedema fluid can also build up in body cavities such as the abdomen, when it is referred to as ascites. The condition may be due either to low levels of circulating protein in the blood, or to heart failure and a failing circulation. In Angora goat kids, a condition referred to as 'swelling disease' is occasionally reported, and linked to vitamin E deficiency; profound anaemia is a feature of the condition (*see* Chapter 12).

Jaundice

Like anaemia, jaundice is not a disease in itself, but is a clinical diagnosis that occurs when too much of a substance called bilirubin builds up in the bloodstream. There is a low background level of bilirubin in healthy goats that is produced during the normal recycling process of red cells/haemoglobin. Jaundice is an indication that these normal background levels have been exceeded, with potentially adverse results. Jaundice can only be detected clinically when severe, when the visible mucous membranes (conjunctiva of the eye and lining of the mouth) show a yellow/brown colour. There will usually be other, more generalized signs, such as dullness, depression or diarrhoea, depending on the cause of the jaundice.

Blood samples are able to pick up early cases of jaundice, and differentiate between the two main causes. It is also important that any death in which jaundice was a feature clinically when the goat was alive, is subjected to a post-mortem examination to ensure that any other cases that may develop can be treated correctly (or more importantly, prevented). The two main causes are:

- 'Haemolytic': when there is massive destruction of the red cell series. Causes include copper poisoning (*see* Chapter 16), poisoning by certain plants including kale, rape or other brassicas, and certain infections including leptospirosis.
- Liver related: usually as a result of obstruction of the bile ducts by liver fluke (*see* Chapter 7), a tumour, or an abscess.

DISEASES OF THE LYMPHATIC SYSTEM

The most important disease affecting the lymphatic system worldwide is undoubtedly caseous lymphadenitis (CLA).

Caseous Lymphadenitis (CLA)

Cause: The bacterial organism *Corynebacterium pseudotuberculosis*, so-called because the abscesses produced in the lymph glands can resemble true tuberculosis (which in reality is rare in goats). The same organism can cause a similar disease in sheep, and goats can pick up infection from sheep and vice versa.

Clinical presentation: The organism becomes established in the lymph glands in almost any location within the body, but affected glands are mainly superficial. The diagram on p. 108 shows the most commonly affected glands in goats. These glands may not be readily found unless they are enlarged, when they become progressively more obvious, before eventually bursting, liberating pus. Internal lymph glands can occasionally become infected (for example, the thoracic or abdominal lymph nodes), with abscesses also developing in internal organs such as the liver, lungs or kidneys. Lesions have also been described in the lymph gland draining the udder (supra-mammary gland): if an abscess bursts, the ensuing pus could potentially infect the milk.

This Boer goat has a well-established CLA abscess in two lymph nodes.

or by taking a swab from a burst abscess. There are blood tests available in some countries, though many are only valid for sheep; these tests can give variable results and need careful interpretation.

Is it infectious? The organism is spread mainly via pus escaping from a burst abscess, either directly – one goat brushing up against an infected goat – or indirectly (pus getting on to, for example, a pen fixture such as a food trough, from which it can be picked up by another goat). The organism gains entry through an abrasion or other break in the skin, from where it is carried via the lymphatic system to the nearest lymph gland that will then become infected. Until the gland bursts, the goat is not too infectious, the pus literally being contained within the abscess wall. However, the pus itself is very resistant in the environment.

All age groups of goat can be infected.

Treatment: There is no reliable treatment available, since antibiotics have a very limited effect. There may be a temptation to lance and treat an infected gland by irrigation, but it is likely that other glands may be affected elsewhere, so the effect is only transitory; you also run the risk of transmitting infection to

The condition is rarely fatal, but if the internal organs are affected, then a variety of clinical signs can develop, depending on the location, size and number of abscesses involved.

Confirmation of disease: The clinical signs are fairly typical, but confirmation of infection can be made by laboratory examination of either pus aspirated via a syringe and needle,

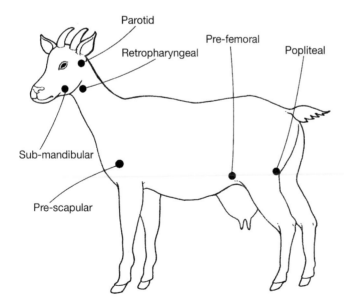

The location of the superficial lymph nodes most commonly affected with CLA.

others. Management of an outbreak depends to a large extent on the type of unit. Large herds will often cull quite heavily to prevent infection spreading, while on smaller units with pet goats the problem is 'managed' by regular observation and handling of goats. If a swollen gland is identified, then it may be better to keep the goat in isolation until the abscess bursts (or is lanced: discuss this approach with your vet); the goat can then be re-introduced to the herd after the wound has healed.

Prevention: The best means of control is by keeping infection out! Any purchased goat is a potential risk unless you know that the herd of origin is free of infection. Avoid buying a goat with swollen glands in any of the sites shown in the diagram. Furthermore, as an apparently healthy goat may also be incubating disease, it is important to keep purchased goats in quarantine away from the main herd for at least two months, observing and handling regularly for signs of a swollen gland. This approach will allow an abscess to develop 'safely'. Do not forget that infected sheep can also transmit disease, and that infection can be carried on equipment such as feed troughs and weigh crates: if you are borrowing or buying these, make sure they are thoroughly cleaned and disinfected before use.

Tumours

Cancer of the lymphatic system and lymph cells (lymphocytes) can occur. Thymomas have already been described in Chapter 8. A more generalized form, in which all body lymph nodes become progressively enlarged, is referred to as lymphosarcoma. Your veterinary surgeon may diagnose this by taking a blood sample or a lymph node biopsy. There is no cure.

CHAPTER TEN

The Nervous System

The nervous system of a goat is made up from two main systems: the central nervous system, commonly referred to as the CNS and consisting of the brain and spinal cord; and the peripheral nervous system, essentially those nerves travelling from the CNS to the remainder of the body. The majority of the problems encountered are related to disease or abnormality of the CNS.

The main parts of the brain are the cerebrum, or fore brain, made up of the right and left cerebral hemispheres; the cerebellum, or hind brain; and the brain stem.

The cerebrum: This is the largest area of the brain and is concerned with all higher mental functions, such as, in humans, thinking and memory – although functioning at a lower level in the average goat, of course! It is made up of two halves, or hemispheres: the right cerebral hemisphere controls the left side of the body, and the left cerebral hemisphere controls the right side of the body.

The cerebellum: This is the hind part of the brain and is concerned with balance and co-ordination. These activities are carried out automatically (subconsciously) by this area of the brain, and are not under the goat's control.

The brain stem: The brain stem controls the basic functions essential to maintaining life, including blood pressure, breathing, heart-beat and also eye movements and swallowing. It is at the base of the brain and connects the cerebral hemispheres to the spinal cord. A number of important nerves called the cranial nerves have their roots in this area (important in the development of listeriosis, *see* later in this chapter).

The entire brain is contained within the bony skull. Any swelling, such as an abscess, haemorrhage, tumour or parasite cyst, is potentially serious and life-threatening, because the resulting increase in pressure causes damage to the underlying brain.

The spinal cord: This extends down from the brain, through the vertebral column encased in, and protected by, the bony vertebrae. Nerves leave the spinal cord in pairs along its length, supplying the skin muscles, bones and other body organs (this is the peripheral nervous system).

Neurological diseases – that is, diseases affecting the nervous system – are quite common in the goat, with virtually any age affected, from a newborn kid to an elderly goat. They can be very difficult to diagnose, and your vet will need to carry out a detailed examination to decide on the cause. Part of this examination will be obtaining a good history, and this will help your vet find out what may be wrong; the sort of information he would probably require might be as follows:

- What is the age of the goat?
- Is this an isolated case, or was more than one case seen?

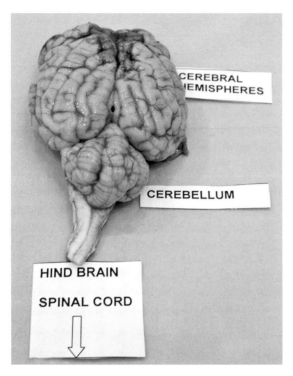

A normal goat's brain, with the right and left cerebral hemispheres, the rounded cerebellum and the upper part of the spinal cord.

- The duration of the clinical signs: did the problem develop rapidly, or over a few days?
- What signs have you observed?
  - Blindness?
  - Deafness?
  - Drooping of ears/eyelids?
  - Dropping food from the mouth (quidding)?
  - Hyperaesthesia – very jumpy and nervous when touched, or over-reacting to a noise or other stimulus?
  - Circling?
  - Fits or convulsions?
  - Head pressing – literally standing with the head pressed against a wall, often in a corner on its own?
  - Dull and depressed?
  - Hind-limb weakness?
  - Collapsing?
  - Coma?

All these signs can indicate an abnormality of the nervous system, but many can also be seen in other conditions, such as some poisonings (*see* Chapter 16) or metabolic problems such as hypocalcaemia (*see* Chapter 5).

NEUROLOGICAL PROBLEMS IN NEWBORN KIDS

The following diseases are those that would be evident at the birth of the kid.

Swayback
Cause: A deficiency of copper in the diet of the pregnant doe, or blockage of copper uptake from the diet due to an interaction between copper, molybdenum and sulphur in the diet. It is important that a well-balanced ration is fed.

Clinical signs: A variety of signs may be seen, depending on the severity of the damage that has occurred to the CNS during the development of the kid in the womb. In mild cases, newborn kids are bright and alert and able to suck, but may show signs of weakness in the hind end; in more severe cases the kid will be totally unable to support its weight. If the damage is more severe, and the brain is affected, then kids will be profoundly dull, may develop fits, and will die quickly. Affected kids may be unable to stand and suckle, and may not consume adequate colostrum, thus increasing their susceptibility to hypothermia and infection (*see* Chapter 6). Swayback may also affect kids up to several weeks of age – so called 'delayed swayback', *see* later in this chapter.

Treatment: There is no specific treatment. The quick recognition of affected kids is essential, and some kids may survive and grow through the problem if they are nursed (given colostrum by stomach tube if they haven't fed), kept warm, and turned over regularly if they are recumbent. Severely affected kids may have to be destroyed humanely.

Prevention: This is based on ensuring that the copper status of does is maintained during the critical period of mid-pregnancy, when the kid's nervous system is being formed. Copper has a vital role to play in the formation of myelin in the developing foetus; myelin itself is the fatty covering of nerve fibres, damage to which is the underlying cause of sway-back. There are a number of different ways of achieving this, using drenches, boluses or injectable copper preparations. Copper is potentially toxic, however, and great care must be taken when considering copper supplementation, as too much copper will kill a goat (*see* Chapter 16), and the balance must be carefully maintained. Your vet may well suggest checking the copper status of does during pregnancy before advising on the best course of action.

Border Disease

The potential role of this virus as a cause of abortion and neonatal weakness has been described earlier in the book. In the UK, it is a much more commonly encountered problem in sheep flocks, and it is this fact that may occasionally lead to problems in goats if the species are kept together, although the disease is not common.

Cause: A virus that can be carried by apparently healthy sheep and goats, and introduced quite unknowingly into a group of susceptible goats.

Main signs: The virus has an ability to cross the placenta of a susceptible pregnant goat and cause damage to the developing kid (it is in the same family of viruses as German measles in humans, but is distinctly different, and is not a risk to human health). The degree and type of damage to the foetus that occurs depends on the stage of pregnancy. Some does may simply lose their foetuses very early in pregnancy (so-called embryonic loss) resulting in a barren doe; later pregnancy foetuses may be aborted. The primary target in more developed foetuses however is the nervous system, and affected kids born at full term may show a variety of neurological problems as a result. The clinical presentation may vary from a kid that shows a fine muscular tremor, to one that

is constantly convulsing due to an absence of its cerebellum, literally destroyed by the invading virus. Some kids may, however, be born normally, and survive to adulthood, but have themselves become virus carriers and will therefore perpetuate the virus transmission to more susceptible goats.

Treatment: None. Severely affected kids must be destroyed on humane grounds. Less severely handicapped kids may survive with good nursing.

Prevention: Infection is introduced by the purchase of apparently healthy goats (or, more commonly, purchased sheep). The condition is unusual in goats, but on both the occasions that it has been encountered by the author, the source of the virus was thought to be sheep that were co-grazing. It is possible to test for the presence of the virus by a blood test – discuss this with your vet.

Brain Damage/Hypoxia
See Chapter 6.

Floppy Kid Syndrome
See Chapter 6.

NERVOUS DISEASES THAT DEVELOP IN KIDS NORMAL AT BIRTH

Delayed Swayback (Enzootic Ataxia)
This condition has already been described earlier in this chapter, as it affects newborn kids. The condition may also develop in older kids up to several weeks or even months of age.

Main signs: Affected kids are normally bright and alert in the early stages of life, but progressively fine muscle tremors will develop with head shaking, quickly progressing to hind-limb weakness (termed 'ataxia'), and paralysis of the hind end. If the whole herd is copper deficient, then a variety of other signs may also be apparent including decreased milk production, anaemia, loss of coat pigment colour, and poor fleece quality in fibre goats.

The bone density of young growing kids may also be reduced, with an increased susceptibility to fractures. Your veterinary surgeon should be consulted if any of these signs are seen in combination, and blood samples can be taken to check on the copper status.

Treatment: None. Severely affected kids should be destroyed on humane grounds; less severely affected kids may survive with good nursing. The remainder of the herd may benefit from copper supplementation if other signs are present, but discuss this with your vet as copper is potentially poisonous if given to excess.

Prevention: This is based on ensuring that the copper status of does is maintained during the critical period of mid-pregnancy, when the kid's nervous system is being formed.

Disbudding Meningitis/Encephalitis

Cause: Disbudding is a skilled procedure, carried out on a 'high risk' patient. Occasionally damage can occur to the underlying brain, causing either immediate physical damage to the underlying brain (pressure or heat), or allowing secondary bacteria to enter the cranial cavity and cause infection (*see* Chapter 6).

Main signs: These vary depending on the severity. Some kids may be found dead, others show signs varying from depression and dullness, to fitting and convulsions. These signs may be seen immediately after disbudding, but may not be apparent until several days later.

Treatment: Consult your vet; severely affected kids may need to be destroyed on humane grounds, those less severely affected may respond to antibiotics and nursing.

Bacterial Meningitis

Cause: This is a bacterial infection of the brain and its covering membranes or meninges. It usually develops as a result of kids being born into a dirty environment, and is a particular problem in kids consuming insufficient colostrum.

Age group affected: Kids from a few days to a few weeks old.

Main signs: These are variable, and include lethargy and failure to feed, compulsive wandering and head pressing, leading to convulsions and death. There may be other clinical signs including diarrhoea and navel ill, and swollen joints.

Treatment: Must be rapid to be successful. Consult your vet; antibiotics will be prescribed, and nursing is important.

Prevention: Ensure that your doe kids in a clean environment, dip the navel immediately after birth, and ensure that adequate colostrum is consumed.

Spinal Abscess

Cause: Bacteria localizing in the spinal canal, usually as a result of navel infection.

Age group affected: Kids from a few days to a few weeks old.

Main signs: A gradual onset of hind-limb weakness, although the exact signs seen will vary depending on the location of the abscess in the spinal canal, where it causes pressure on the spinal cord and associated nerves. Kids may completely lose the use of their hind limbs as the pressure builds. The signs can be confused with swayback, so seek advice from your vet.

Treatment: In the early stages, antibiotics may be beneficial, but once established, treatment is usually disappointing, and affected kids may need to be destroyed on humane grounds.

Prevention: Ensure that your doe kids in a clean environment, dip the navels immediately after birth, and ensure that adequate colostrum is consumed.

Trauma

Young kids are adventurous and inquisitive, and may well become injured as a result. They may also be kicked, butted or trodden on by bigger/older goats, and suffer concussion or spinal damage as a result.

Tetanus

Cause: A bacterial infection caused by *Clostridium tetani*. It is an organism found fairly commonly in soil, particularly on farms with livestock. Infection is caused by spores of the organism gaining entry to a wound (a cut or puncture wound); infection may also follow disbudding, ear tagging or tattooing if these procedures are undertaken unhygienically. Once established in the wound, the spores become activated, releasing a powerful toxin causing damage to the nervous system.

Age group affected: Almost any age of goat can become affected, but it is most likely to be seen in young kids.

Main signs: There is a gradual onset of stiffness that progressively affects the whole body. The ears are held erect, the tail is raised up, and the limbs are held out stiffly because the relevant muscle groups become permanently contracted, leading to bouts of muscular spasm, particularly when the animal is disturbed by sound or touch. As the disease develops, the goat becomes recumbent on its side, with its legs held out rigidly. Most affected goats die of the condition.

Treatment: Very disappointing and rarely effective. Affected goats may die a horrible death, and euthanasia should be considered if the condition continues to deteriorate. Some goats will pull through if it is recognized early, and the goat nursed intensively.

Prevention: There is an effective vaccine available, and does should be vaccinated/boosted with an appropriate clostridial vaccine in late pregnancy, thus ensuring transfer of immunity via the doe's colostrum to vulnerable kids. Ensure that all invasive procedures such as disbudding, castration, tagging and so on are all done in a clean, hygienic manner, and that kids are returned to a clean environment.

NEUROLOGICAL PROBLEMS IN GROWING KIDS

Trauma
See earlier in this chapter.

Tetanus
See earlier in this chapter.

Cerebrocortical Necrosis (CCN)
This disease is also referred to as polioencephalomalacia.

Cause: A deficiency of vitamin B1 (thiamine). In normal, healthy goats, this vitamin is produced in the rumen by the micro-organisms that live there, and deficiency is normally related to an interruption of this manufacturing process. It normally follows an increase in the number of certain bacteria capable of producing a thiaminase enzyme, that literally destroys thiamine before it is absorbed. This can occur following an abrupt change in diet, or particularly following the ingestion of mouldy feed or hay.

Age group affected: The goat must have a functional rumen, and although young growing goats are most susceptible, the problem can occur at almost any age.

Main signs: Dullness initially, often accompanied by mild diarrhoea. As the condition progresses, the goat becomes blind, and has difficulty standing as it appears to want to hold its head up and back (a condition often referred to as 'star-gazing'). It then moves to lateral recumbency (lying on its side) with its legs paddling, and its head thrown back apparently convulsing – this is referred to as 'opisthotonus'. Untreated animals will die rapidly, and veterinary attention should be sought early in the course of the disease.

Treatment: Your veterinary surgeon will decide initially if CCN is the underlying problem. If it is, then the administration of thiamine in high doses intravenously can bring about a spectacular response. The goat must be treated in the very early stages to be successful, however, and, as with most nervous disease problems, nursing is vitally important.

Prevention: As the cause is not really known, it is difficult to be too precise. If CCN is confirmed, however, then you should consider what may have triggered the problem.

Caprine Arthritis Encephalitis (CAE)

Cause: A lentivirus, a virus that causes a wide range of clinical signs (mainly in older goats), and is characterized by its slowly developing clinical picture (*see* Chapter 11).

Age of goat affected: The encephalitic form of CAE is rare, but will be encountered occasionally in goat kids aged 2–4 months, that were exposed to the virus in early life.

Main signs: Initial signs include an apparent lameness or stiff gait, followed by ataxia (unsteadiness). Affected kids progressively lose the use of their limbs, and this disability may be in all four limbs, two front or two back, or even two left limbs or two right limbs. This progresses to recumbency, with obvious involvement of brain, with convulsions and blindness developing. The disease can be confirmed accurately only by post-mortem examination, but a blood sample will prove that the goat kid carries the CAE virus.

Treatment: There is no treatment, and severely affected kids may need to be destroyed on humane grounds.

Prevention: The control of CAE is complex: refer to Chapter 11.

Lead Poisoning

Young goats are inquisitive, and may consume anything interesting or unusual (*see* Chapter 16).

Louping Ill

Cause: A virus transmitted by ticks, and confined geographically only to tick areas.

Age group affected: This disease can be seen in any age group, but young kids on units where louping ill is a regular problem will be strongly immune in their first year if they consume sufficient colostrum from an immune dam; however, they will be susceptible to infection in their second season. Conversely, if a dam is sick with louping ill (meaning that she hasn't yet produced antibody), her milk will be infected, and will be transmitted to a suckling kid.

Main signs: An initial lethargy, followed by intermittent head shaking with twitching of lips, nostrils and ears. Then generalized muscle tremors, muscle rigidity and stiff jerky movements follow these early signs. Finally there is a loss of balance, recumbency and death.

Treatment: None available. Your veterinary surgeon will decide if your goat may benefit from nursing and supportive treatment, but may advise euthanasia on humane grounds.

Prevention: This disease is spread by ticks and is also a problem in sheep, which are rapid multipliers of infection by re-infecting tick populations. Basic control is aimed at recognizing those parts of your farm or grazing area that may harbour ticks (usually rough hill grazing), and either avoiding it, or trying to improve the habitat. Your vet may advise the use of a product to control ticks on your goats, and there is also a vaccine available.

NEUROLOGICAL PROBLEMS IN ADULT GOATS

Listeriosis

Cause: A bacterium *Listeria monocytogenes*, widely found in the environment around many farm units, particularly in stagnant water, muddy puddles and rotting vegetation. The organism can also multiply rapidly in poorly made or maintained silage, which is the single most important source of disease.

Age group affected: The condition can affect goats of any age, but the neurological form most commonly affects young adult goats or older goats that may be losing their teeth. This is because the infection is thought to get into the brain from the mouth, in particular via damage to gums or the cheek by erupting teeth, tooth loss or abnormal tooth wear. The bacteria are then thought to track along the local nerve fibres in the mouth, eventually moving to the brain along cranial nerves. Most cases of disease are seen in housed, silage-fed goats. Sporadic cases can occur when goats are outside; one case encountered by the author involved disease in two adult goats

This clamp was poorly sited, and contained poorly made silage; levels of listeria were very high, resulting in a serious outbreak of disease.

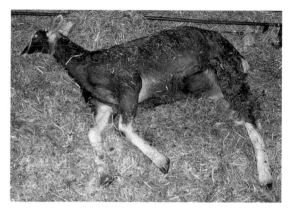

This goat is at a later stage (refer to colour section for goat at an earlier stage); it is now lying 'flat out' and is unable to sit up.

that had been browsing around a heap of old horse manure (a source of *Listeria* organisms), covered by thistles and brambles, both capable of causing damage to the inside of the mouth when eaten, thus allowing the organisms to enter the damaged skin.

Main signs: Disease can present in a number of different ways. In typical *Listeria encephalitis*, the early signs include dullness, depression and inappetance. As the condition develops, signs of facial paralysis begin to show, such as the ears and eyelids drooping, the tongue protruding, and drooling saliva. Eating becomes progressively more difficult, and cudding becomes haphazard, with bits of food dropped from the mouth while chewing. As the condition progresses still further, there may be a noticeable head tilt, and a constant sideways motion of the eyes or 'nystagmus' may be noticed. The goat may walk around in circles, often with a stumbling awkward gait, or simply stand with its head pressed up against the wall. As the condition progresses (and this whole set of clinical signs can occupy only a few hours in severe cases), the next stages will be recumbency with early convulsions, followed by coma and death. The mortality rate, even when recognized and treated, can be as high as 30 per cent.

Not all listeria cases develop in this way; in some cases the timescale can be extremely short, and an affected goat may simply be found dead. Listeria abortion has been described in Chapter 5.

Treatment: Consult your vet; any delay in getting the correct treatment can be fatal. The earlier an infected goat receives treatment, the more effective it is likely to be. High doses of antibiotic are required (intravenously initially), and fluids are vitally important, as many goats cannot swallow, and can therefore rapidly become dehydrated, and develop renal failure as a result.

Prevention: As already stated, many cases of listeriosis result from the feeding of poor silage. The listeria population originates from soil contamination during the ensiling process, and this occurs most commonly when there are molehills or other earth workings in the field when the grass or crop is being cut and gathered. The principle of silage making is to allow forage to mature and ferment in an air-free environment to prevent spoilage. It is important therefore to ensure that the clamp is properly sealed, or that bags or wrappings used for baled silage are of good quality – the aim is to keep air out as the silage matures. Do not feed visibly spoiled silage to goats, and remove uneaten silage after twenty-four hours as it can rapidly spoil, with a build-up of listeria occurring as a result.

Scrapie

Cause: This is a highly topical and important problem worldwide. Scrapie is a member of the group of diseases referred to as 'TSEs' or

This silage clamp on a large commercial goat unit uses tyres to hold down the sheeting, thus keeping air out.

'transmissible spongiform encephalopathies'. The group also includes the cattle condition BSE, and more importantly the human condition referred to as vCJD (variant Creutzfeldt Jacob Disease). The group of diseases is caused by infectious agents referred to as 'prions', which are unlike either viruses or bacteria as they do not contain DNA. There are a number of separate strains of prion, which together with host genetic resistance and susceptibility factors, lead to the range of clinical presentations described.

Scrapie has affected goats for many years and, together with scrapie in sheep, is a well-documented problem. Recently there has been much interest in the development of scrapie-resistant sheep, linked to the recognition of varying resistance patterns genetically, and a drive to breed in resistance, and conversely eradicate the susceptible sheep lines. Sadly this is not the case in goats, and although some resistance has been recognized experimentally, the procedures are not robust enough to develop control programmes; thus at the time this book was written, one confirmed case of scrapie in an EU goat herd will result in whole herd slaughter of the remaining animals.

There was a concern for some time that goats might have picked up the BSE agent, as many were fed the same feed in the 1980s and 1990s that contained infected meat and bone meal, the source of infection in the cattle epidemic. In 2005 it was revealed that a goat in France was confirmed with BSE, having been picked up in a screening exercise in 2002, and a similar conclusion (that is, BSE infection) was made following the re-examination of the brain of a goat that died in Scotland in 1990. Nevertheless, scrapie remains a rare, albeit important neurological problem of goats – and it is also a notifiable disease.

Age group affected: Scrapie usually affects goats over about two years of age.

Main signs: As with disease in sheep, scrapie presents in two forms with considerable overlap between the two – neurological and pruritic (itchy skin). The disease is insidious, and may take many months to develop, early signs being very subtle and often missed. In the early stages, the goat may appear apprehensive and excitable, become unsteady on its hind legs, and show fine muscle tremors over the body and head. Its appetite will be reduced, its milk yield will decline, and it will begin to lose weight. Pruritus will progressively develop, with nibbling of its limbs and a constant rubbing against any solid object – touching will often provoke a violent itching response. The nervous signs become progressively more severe, with hind limb incoordination and a peculiar high stepping gait of the fore limbs. Deterioration can be rapid and terminal: death is inevitable.

Confirmation is based on an examination of the brain tissue, looking either for vacuolation in the brain, or other more specific changes.

Treatment: There is *no* treatment. The disease is notifiable, so if you suspect that your goat is infected with scrapie, contact your own vet – but both you and your vet have an obligation to report this suspicion to the relevant national veterinary authority (Defra in the UK). There is compulsory slaughter with compensation paid.

Prevention: As already stated, there is no known genetic resistance in goats that can be utilized to breed resistant goats, although research is on-going. If a case is identified, then whole herd slaughter is a likely scenario.

A brain showing typical scrapie vacuole (hole) formation, under high magnification.

Louping Ill
See earlier in this chapter.

Pregnancy Toxaemia
See Chapter 5.

Hypocalcaemia
See Chapter 5.

Hypomagnesaemia
Cause: A lowered level of magnesium in the blood: a goat is unable to store magnesium and needs a regular daily intake. It is a rare condition in the goat, but may be seen in high-yielding goats grazing on lush, wet herbage, and is only likely to occur on small units as a result. Goats on larger units are most often housed, and fed a well-balanced diet.

Main signs: Excitement, tremors, over-reaction to stimuli (hyperaesthesia), then collapse followed by convulsions and death. The course of deterioration can be very rapid, and goats may just be found dead. A milder form can occur, in which goats are inappetant, apprehensive, go off their feed, and suffer a drop in milk yield.

Treatment: Acute cases require rapid treatment: call your vet as a matter of urgency. Your vet will normally administer magnesium (usually with calcium) intravenously –

this is a skilled job, *so do not attempt it yourself* as the goat may die if magnesium is given too rapidly. Your vet may advise that you give magnesium sulphate solution (if you have some) sub-cutaneously if there is likely to be a delay before his/her arrival.

Prevention: Provide a well-balanced diet, with plenty of fibre (hay/straw/hedge cuttings). Ensure that goats outdoors have shelter.

Coenuriasis (Gid)
Cause: The cyst of a tapeworm, the adult form of which (*Taenia multiceps*) lives in the gut of dogs and foxes. Segments of the tapeworm containing eggs pass out in dog faeces on to the pasture, and can potentially be picked up by grazing goats (or sheep). The eggs hatch in the intestine, and then move via the bloodstream to the brain. Once in the brain, usually a single cyst (referred to as a *Coenuris cerebralis* or a 'Gid cyst') grows slowly over several months, producing gradually worsening signs. If untreated the affected goat or sheep will die. The life cycle is perpetuated if a dog or fox then eats the goat or sheep head. In the UK the condition is more commonly seen in sheep-rearing areas, particularly in Wales. If goats are affected, it is normally because they are kept in a sheep area with a greater chance of dog tapeworm infestation.

Age group affected: Although comparatively rare in goats, the condition will be seen mainly in growing kids or young adults.

Signs: Depend entirely on the location of the cyst in the brain, with local pressure being exerted on brain tissue as the cyst increases in size. The majority develop in the cerebrum, and early signs include dullness and depression; then as the condition progresses, blindness will develop in one eye only, the head will be tilted to one side, and the goat will be unable to walk in a straight line, and may begin to walk in circles. In some cases, as the pressure builds, the bony covering of the skull may soften. Your veterinary surgeon will carry out a full neurological examination to decide on the cyst location if Gid is suspected.

Treatment: Depending on the location, your vet may be able to carry out brain surgery, and remove the cyst through a small hole drilled into the skull. Those in the cerebellum are usually deemed inoperable due to the damage that has already occurred. Post-operative recovery with cerebral cysts can be very good.

Prevention: This is based on an understanding of the life cycle. Worm, on a regular basis, any dogs that live on your farm or holding with an effective wormer for tapeworms. Ensure that dead goats (or more importantly dead sheep) are not left out in the fields; remove them and dispose of them. Sheep heads have occasionally been fed to farm dogs on sheep farms as a delicacy – something to be avoided!

Traumatic Injuries in Adults

Concussion and spinal injury can occasionally be encountered in adults (particularly bucks) as a result of fighting. If aggressive bucks are kept together, injury can be severe, and may even result in a broken neck. Good buck management will recognize potential conflicts before such injuries develop. A broken neck can result in a buck being found dead. Injuries lower down the spinal column may lead to varying degrees of fore- and hind-limb weakness or paralysis. Consult your vet as a matter of urgency; do not be tempted to move an injured buck for fear of causing more severe injury to the delicate spinal cord. Remove other goats from the pen.

Hind-Limb Nerve Damage/Paralysis

This condition can occasionally develop in does after a difficult kidding (*see* Chapter 5). The delivery of a large kid, or the pressure caused by a kid stuck part way out, may damage nerves (the obturator nerve) to the hind limbs that run on the floor of the pelvis; such damage may cause a goat to walk with 'splayed legs', and on a slippery surface such a goat may literally 'do the splits', potentially causing more damage.

A goat that is recumbent for a number of days can also damage a separate nerve (the peroneal nerve) running down the side of the hind limb. When the goat eventually stands, the damaged nerve may cause the hind limb to knuckle at the fetlock.

These nerve injuries will often recover with good nursing and with the provision of soft, non-slip bedding, and also physiotherapy – consult your vet if you suspect such an injury.

CHAPTER ELEVEN

The Locomotor System

Throughout this book, the importance of protecting goat welfare has been emphasized, to the effect that 'a happy goat is a healthy and productive goat'.

One of the most important causes of pain and discomfort in a goat is lameness, and a lame goat must be recognized quickly, a diagnosis made (there are many causes of lameness), and the correct treatment, including pain relief (if necessary) given rapidly. Almost any type of lameness, if neglected, will not only cause suffering and distress in the affected goat, but may also result in chronic long-term damage. Lameness can result from disease or damage to the feet, bones, joints, muscles or other soft tissue structures of the limbs, and can be caused by developmental or congenital abnormalities, trauma, infection or nutritional imbalance. As with any disease or abnormality affecting a goat, prevention should always be the primary goal, and preventative measures will be emphasized throughout this chapter. Basic preventative measures should include:

- Regular examination of the feet to ensure that overgrowth does not occur.
- Regular examination of the environment to identify and remove potentially hazardous insults (for instance stones, flints, hedge trimmings) that could be caught in the feet as goats move along a route to the field.
- Recognition of those infectious diseases over which you have some control – for example CAE, by testing purchased animals if you are free of disease, to prevent CAE being introduced. Footrot is another infectious disease that can be kept out of a clean herd, by careful attention to herd biosecurity.
- The provision of a good, well-balanced diet will prevent nutritional mineral or trace element imbalance.

IDENTIFYING LAMENESS

No matter whether you have one goat or hundreds of them, the key to maintaining their good health is regular observation. By observing your goats on a daily basis, you will get to know their habits and peculiarities, but more importantly, you will recognize when things are going wrong quickly. If a goat is suddenly lame it will be fairly obvious, it will hold up its leg, and will be reluctant to use the affected limb, or it may begin feeding on its knees – it may even be recumbent, particularly if there is a severe problem such as a fracture. In a progressively worsening lameness, it is important to recognize the early subtle signs of lameness, including a 'limp', or reluctance to come up and feed. Try and build up a rapid picture of what may be wrong: is it the first case of lameness you have seen, or are a number of goats showing signs?

The next step is to catch your goat and examine the limbs (remember that more than one foot may be affected, it is just that the

Although a goat that is severely lame on its front feet may feed like this, some goats prefer to adopt it as their normal stance. Know your goat's behaviour!

goat will normally favour the worst affected limb). Try to catch the goat without chasing it around the pen or field, as you may cause more damage if the goat panics. Most lameness problems affect the feet, so unless there is an obvious injury or swelling further up the leg, start by examining the feet. See below for a diagram of the anatomy of the foot. Unless your goat is used to being tethered, then you will probably need someone to hold the goat's head while you carry out your examination. For this, use the following procedure:

- If the foot is dirty, clean it with a brush and water.
- Initially look between the claws for soreness (known as 'scald'), or for any foreign body such as a stone, thorn or dried ball of mud or faeces – both a stone and a mud ball can cause damage, pain and discomfort to the soft skin between the claws. Look out for under-running of the sole, which may indicate, for example, footrot.
- If there is nothing immediately obvious, then try to work out where the pain is located; it may be worse in one claw, and you can often recognize this either because the affected claw feels warmer (a developing abscess), or the goat flinches when you apply gentle pressure.
- The next step is to begin to gently pare away any loose or under-run horn, or to pare the horn over the white line – but this is a skilled job, and should only be carried out by an experienced goat keeper.
- Contact your veterinary surgeon if you are not sure what the problem is, or if severe damage or injury is identified. Your vet may decide to use antibiotics if infection is identified, or may advise that you walk the goat through a footbath, or stand its foot in a bucket of, for example, formalin, copper sulphate or zinc sulphate.

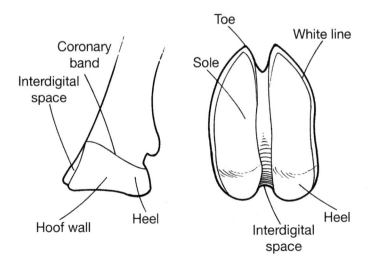

Normal anatomy of the foot.

The foot must be picked up and examined if a goat goes lame.

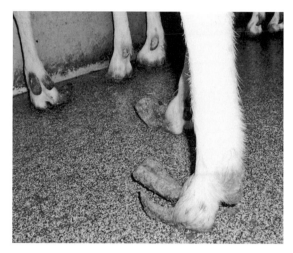

This foot is grossly overgrown and neglected, and is probably due to a genetic predisposition.

Remedial Foot Trimming

The feet of all goats will grow constantly, but will be naturally worn down by the goat's normal movements, particularly if moving over hard or rough surfaces. The feet of some goats will have a tendency to overgrow however, either because of the shape of the foot (often genetic), or because they are kept on soft surfaces which prevent normal wear and tear – this is a potential problem in groups of goats housed all year round on deep litter. This overgrowth is further exacerbated in commercial goats by heavy concentrate feeding that may predispose to laminitis and foot overgrowth. As a result, it is good management practice to examine the feet of all goats at least two to three times each year. This does not mean you have to trim them every time you examine them, however, only if they are overgrown!

The basic equipment that you will require includes a pair of small secateurs (footrot shears/sheep secateurs), and a small double-edged hoof-knife with a curled toe is also useful, although many shepherds use a simple but sharp, straight-edged knife.

It is best to work with the goat standing, haltered and tied short so that only minimal evasive movement is possible. There should be a wall or similar obstruction available, so that it cannot circle away from the operator during the trimming procedure. Ideally tie it in a corner between two walls, so that it may be properly restrained whichever side the operator is working. Overgrowth problems that may be identified include:

- Heels too high – tipping the bodyweight forward, causing undue wear at the toe, and producing a very straight leg; the outer heel may be higher than the inner heel.
- The outer claw tends to grow much longer

> **WARNING!**
>
> In unskilled hands, a foot knife, clippers or shears can have devastating consequences: by cutting too deep, or in the wrong place, excessive bleeding can occur, and the delicate, deeper tissues of the foot can be exposed and permanently damaged. This applies equally to paring the foot of a lame goat, or remedial (preventative) foot trimming – *see* below.
>
> Ask your vet to show you how to examine the feet, and how to investigate lameness and trim the feet. You may find that there are local sheep discussion groups or agricultural colleges where training courses may be held; or ask a skilled shepherd to help.

Front view

Side view

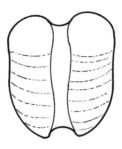
Solar view

The ideal foot shape to aim for if paring misshapen feet.

than the inner claw, and may then either grow across the inner claw, or turn up at the end like a Turkish slipper. If left untrimmed, this overgrowth causes progressive strain on the flexor tendons of the foot.

- The outer wall may turn inwards and grow under the sole (that is, over the weight-bearing surface) such that the space created fills with grit and earth causing abrasion, deeper damage and infection.
- The inner walls of each claw may occasionally grow across the interdigital cleft, increasing the chance of mud or faeces becoming trapped between the claws.

As already stated, foot trimming should only be undertaken if necessary, and not as a routine measure, but the feet must be examined for evidence of overgrowth on a regular basis. Trimming should only be carried out by a trained and/or experienced individual, correcting the abnormalities described above to ensure that normal foot shape is restored, as shown in the diagram.

COMMON CAUSES OF NON-INFECTIOUS FOOT LAMENESS

Most of the common causes of lameness affecting the feet in goats are almost identical to those encountered in sheep, and the reader wanting more information may wish to consult one of many texts written on the subject. The causes of lameness can be divided into two catego-ries, infectious and non-infectious; first we will consider non-infectious causes of lameness.

White Line Disease

The white line is located on the underside of the foot, and is essentially the junction between the sole and the wall, and an area of natural weakness in the foot. As such, it is susceptible to damage by foreign bodies such as small stones or flints, and any damage that develops can quickly fill with mud, gradually becoming impacted, which might lead to deep infection. Wet conditions underfoot may soften the horn, making such damage more likely. The resultant infection can track up the wall of the foot, eventually bursting out at the coronary band. Early damage will be picked up at regular foot trimming.

In early cases the next step is to pare away the horn of the wall in a 'half moon' shape to prevent material becoming trapped. Check

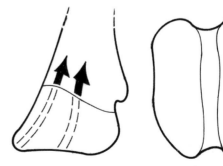

White line infection; watch out for black marks in the white line area, as infection may track up and burst at the coronary band.

along the white line by paring, as there may be other similarly affected areas. If infection has developed, you may find that pus begins to ooze from the area you are exploring, which must then be exposed, taking care not to cut too deeply. If the foot is painful, and you suspect that there may be a more deep-seated abscess (or there is a swelling developing at the coronary band), then it may be better to poultice the foot, using a ready-made poultice from your vet, or a home-made poultice using bread or bran, dampened with warm water and some salt, with the foot and poultice inside a polythene bag taped to the foot/leg.

This is a potentially serious problem; if in doubt, discuss its management with your vet.

Foreign Body Penetration

It is possible for pieces of wire, thorns, sharp stones or flints to penetrate the feet, and some (particularly thorns) can remain embedded and clearly visible when the foot is examined. The goat is usually suddenly lame. The foreign body, if still present, must be removed, but in all cases it is important to open up the site of penetration to allow drainage of pus, as infection is a common sequel. Discuss this with your vet.

These hedge trimmings were picked up after a mechanical hedge trimmer had just passed by!

Laminitis

This condition is normally related to lameness/tenderness in more than one foot, and is essentially a metabolic disorder resulting in damage to the sensitive 'laminae', the vascular tissue lying between the wall of the foot and the sensitive structures contained within the foot. It is similar to the very sensitive tissue found under our finger and toenails, which is so painful when you tear a nail!

Laminitis is usually referred to as acute, sub-acute or chronic. Acute or sub-acute laminitis tends to occur after any severe toxic condition such as mastitis, metritis or retained foetal membranes. It can also occur as a sequel to rumen acidosis (*see* Chapter 7) following the ingestion of abnormally high levels of, for example, barley. Chronic laminitis then follows on some weeks later, usually as a result of repetitive attacks of acute or sub-acute disease.

Clinical signs: In acute laminitis there is a sudden onset of tenderness and discomfort that may just affect the front feet; in severe cases it may affect all four feet. The goat is very disinclined to move, it may become recumbent, or move around on its knees. There are obvious signs of pain and discomfort, with tooth grinding, loss of appetite and a drop in milk yield. The affected feet may be hot to the touch. In sub-acute laminitis, the signs are simply less severe. Chronic laminitis is probably under-diagnosed in larger commercial herds, and may only become apparent when abnormal hoof development occurs, often with deep annular grooves in the wall, and a grossly thickened sole and wall (so-called 'platform soles'). Affected goats may have a peculiar goose-stepping gait, and require regular remedial foot trimming.

INFECTIOUS CAUSES OF FOOT LAMENESS

The second category of reasons for lameness in goats embraces infectious causes of lameness.

Scald

Cause: A bacterium called *Fusobacterium necrophorum*, an organism that is commonly found in the environment, being particularly prevalent on fields that have been previously grazed by goats, sheep or cattle. The organism is able to pass through the skin between the claws and multiply if the foot is excessively wet; it is therefore a more common condition in grazing than in housed goats.

Signs: Lesions are superficial and confined to the skin between the digits; the horn itself is unaffected. It can, however, cause severe lameness in one or more limbs. There can be considerable tissue swelling, the lesion has a fairly characteristic odour, and is often a whitish colour.

Treatment: Individual animals are best treated with an antibiotic aerosol spray obtainable from your veterinary surgeon. If a large number of goats are affected, then it may be easier to walk them through a weak formalin footbath (3 per cent maximum strength). It is preferable to keep the feet dry, which will aid healing; this may mean moving affected goats to a different location.

Prevention: This is problematic, if goats are to be grazed outdoors in wet humid weather, as the organism can be widespread, and the circumstances will be ideal for disease to develop. Regular foot bathing can help, and it is beneficial to try and avoid having areas where goats may congregate, thus contaminating and churning up the ground. Move feed and water areas regularly, for example.

Footrot

Cause: *Fusobacterium necrophorum* and *Dichelobacter nodosus*, working in combination. Footrot can be considered as a more serious complication of scald, with the combination of organisms leading to under-running of the horn, and deep sepsis and necrosis of the underlying tissue of the foot. Whereas *Fusobacterium necrophorum* is widely distributed, *Dichelobacter nodosus* is a true infectious disease, spreading from goat to goat, and introduced into a herd by the purchase of

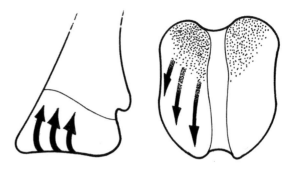

Footrot: infection spreads from the space between the claws and causes separation of the horn of the sole starting near the heel, extending across the sole, and up the wall in severe cases.

infected animals – infection is shared between sheep and goats, so bear this in mind if you have both species! The organism can survive for up to three weeks away from the goat's foot on pasture or bedding, and this is important when attempting to control disease.

Signs: As already stated, footrot begins with damage to the interdigital cleft (scald), the damaged tissue then allows *Dichelobacter* to invade the foot, and the damage is severe, since this organism produces enzymes that literally digest the horn of the foot, causing the horn to become undermined. The lesions tend to begin near the heels, and gradually spread; in severe cases the complete horn can be shed (referred to as 'thimbling'). This is extremely painful, as the delicate and sensitive laminae of the foot become exposed. Disease can be seen in one claw of one foot, or can develop in more than one foot, when the goat will be very lame. As the condition develops, other bacteria invade the damaged tissue, and the whole foot swells and exudes a foul but characteristic odour. The condition can occur at grass, or it may also be encountered in housed goats, particularly if they are overcrowded and the bedding is wet and contaminated.

Treatment: If in doubt, ask your veterinary surgeon to confirm your suspicion of footrot: it is a potentially serious problem and steps need to be taken urgently to treat those affected, and prevent fresh cases. Remember it is a

Footrot.

herd problem, so if you see disease in one goat, expect other cases to follow.

- Examine all feet of all goats, and pare away any loose or obviously under-run horn. Do not be too ambitious, however, and the foot should not be cut so deeply that it bleeds.
- When you have examined all goats, you should be able to split them into an infected and an uninfected group.
- If only a small number of goats are affected and lesions are not too severe, then it will be easiest to treat these with an antibiotic aerosol spray every day.
- If any badly affected goats are identified (and these can be recognized as the foot

will be badly swollen, or deep infection will be identified), they are best referred to your vet, as they will benefit from more intensive treatment, including injectable antibiotics. Once the infection has been brought under control, then the foot can be pared to remove the damaged and under-run horn. Severely affected goats may need to be destroyed on humane grounds.

- For other uninfected goats, the best approach is to stand them in a footbath containing 10 per cent zinc sulphate for ten minutes, or if you have only a few goats, you may prefer to dip their feet alternately in a bucket. Then stand them on dry concrete for the feet to dry, before putting them back into a clean field or a newly bedded yard (remember the organism can live in the environment for up to three weeks), and then repeat the examination and foot bathing again one and two weeks later. Using this approach, you can get on top of a footrot problem.

Prevention: If your goats are free of footrot, there is no reason why they should not remain free. Sources of infection are goats (or sheep) that have been bought in and are infected, or moving goats into contaminated fields or

This is a footbath used to contain, for example, copper sulphate, zinc sulphate or formalin.

Typical foot and mouth disease vesicles (or blisters).

buildings that have carried infected sheep or goats in the previous three weeks.

Vaccines are available.

Foot and Mouth Disease

Foot and mouth disease causes small vesicles or blisters to develop around the coronary band at the top of the hoof wall, or in the interdigital cleft. There is usually acute lameness in all four feet, and similar lesions may also be seen in the mouth and on the teats. This disease is notifiable (*see* Appendix I), and any suspicion of disease must be reported to the government/state veterinary service.

OTHER CAUSES OF LAMENESS IN ADULT GOATS

One of the more important viral diseases affecting goats worldwide is caprine arthritis encephalitis (CAE), with joint infection being the most common clinical presentation.

Caprine Arthritis Encephalitis (CAE)

Cause: A virus of the family Retroviridae; a virus of the same family causes a similar disease in sheep called Maedi Visna (MV). It is a fairly recently recognized problem, with the first report of disease worldwide being made in 1974, since when it has become established in the main goat-rearing countries of the world. Some countries have control programmes, and have successfully eradicated the disease. Infected goats can be recognized by blood or milk tests designed to recognize the CAE antibody produced in response to infection.

Signs: CAE infection results in a range of clinical signs, some of which have already been discussed. The main clinical presentation is lameness linked to arthritic change in the joints. Other signs include encephalitis in kids (Chapter 10), lung infection (Chapter 8), mastitis and udder change (Chapter 14).

Arthritis is seen in goats of six months of age and older. It tends to be chronic and develops slowly, with the carpal (knee) joints most

A CAE infected goat, showing enlargement of the carpal joints.

commonly affected, followed by, in descending order, the tarsal (hock), stifle, fetlock, neck and hip joints. One joint may be affected, or several joints, and although joint involvement may be acutely painful in one goat, another goat may show very ill-defined signs of mild discomfort including stiffness, difficulty getting up or reluctance to move. Again the joint may be visibly swollen, or show only mild change externally.

Infection can lie dormant in a group of goats for many years before clinical disease develops, and the full spectrum of clinical signs already described may not be seen until a high proportion of goats is infected. If you suspect disease, contact your veterinary surgeon and arrange for laboratory tests to be undertaken.

Treatment: There is no treatment, and currently no vaccine available to prevent disease.

Control: This is based on a full knowledge and understanding of the way infection is picked up and spread within a population of goats, and will also depend on the level of infection in your own herd. Once infected with CAE, a goat remains infected throughout the remainder of its life – not all goats will develop disease during their lifetime, and furthermore many goats can be infected with virus but be unrecognized by the owner, as they remain fit and healthy!

The majority of goats become infected by drinking the milk or colostrum of such an infected doe, the virus reaching a high level in the cells of the udder, many of which are shed into the milk. Any other procedure that allows potential sharing of bodily fluids can result in infection being picked up at almost any age. High-risk procedures include the passage of a newborn kid through the vagina of an infected doe; contact with saliva or respiratory secretions; contact with infected blood – by, for example, the sharing of tattoo equipment or needles between two goats (if one is infected). Spread of infection may also occur in the milking parlour, if the milking cluster is placed on the udder of a clean goat when it still potentially contains infected milk from the previous goat.

The first step in control must be to evaluate the CAE status of any herd under investigation. The CAE status can then be quite simply broken down into three main categories:

1. Free of infection – i.e. naïve
2. Low/moderate level of infection – no clinical disease apparent
3. Heavily infected – clinical cases developing

If the herd is truly naïve, the control is based predominantly on keeping infection out by a programme of testing incoming animals, also by ensuring that adequate quarantine measures are in place, and that all other biosecurity measures referred to in this book are adhered to.

Infected herds (low, moderate or heavy): Control measures should be drawn up depending on the level of infection, the type and size of the unit, the available finances, and so on. Each unit is different, and options may vary from doing nothing and living with the problem, to developing a ruthless test and cull policy. Measures to consider include:

• A whole herd bleed, thus giving a population of goats of known status within the limitations of the tests available, and the number of tests undertaken.

• Strategic testing of, for example, breeding females, yearlings due to breed, goats showing clinical signs, and so on.

• A culling policy, depending on what the unit can tolerate with regard to finance, production requirements and suchlike.

• Snatching kids at birth and rearing them away from the dam environment.

• Avoiding the use of pooled colostrum if there is evidence of CAE in the herd, and using either colostrum supplements, colostrum from known negative donors, feeding cow or sheep colostrum, or feeding pasteurized colostrum. Heat-treated or 'pasteurized' colostrum has been held at 560°C for one hour. This time or temperature must not be exceeded, or the colostral antibody protein could be destroyed.

Osteoarthritis

As goats grow into old age (and many pet goats can survive well into their teens), some of them can develop painful joints caused by osteoarthritis, just as old people. It is less of a problem in commercial herds, however, as most goats do not live long enough.

Signs: These include a gradual stiffness and discomfort developing in one or more joints, most commonly the carpus, hock, elbow and stifle. If only one joint or limb is affected, then lameness will become gradually more noticeable. It is not unusual for more than one joint or limb to be involved, however, and signs may then be more subtle, such as the resting of affected limbs alternately, and spending more time lying down. Affected goats will often develop a characteristic stiff gait as they become increasingly unable to fully extend and flex the joints. In time you will begin to see muscle wastage of the affected limb. X-rays may be a useful diagnostic aid.

Treatment: In the early stages in pet goats, it may be possible to control the pain and discomfort by using anti-inflammatory drugs and analgesics, although this is rarely financially viable in a commercial unit. The welfare of the goat must be a paramount concern, however, and euthanasia on humane grounds considered if the condition worsens.

Prevention: None possible.

Lameness after Injection

Temporary lameness in goats can occur following the injection of any potentially irritant material such as some antibiotics, because of the relatively small muscular masses in the hind limbs. Very occasionally a more permanent lameness can result from damage to deeper nerves such as the sciatic nerve in the upper hind limb. Injection techniques are discussed in Chapter 4.

OTHER CAUSES OF LAMENESS IN YOUNG AND GROWING KIDS

Trauma

Young goats are adventurous and fairly agile, and will readily attempt to clamber into, over, under, through or out from a variety of different obstacles in their environment. As a result, they are extremely susceptible to bruises, sprains, cuts and grazes, but also more serious injuries such as fractures and dislocations.

Fractures occur most commonly in the metacarpal area (knee area), the radius/ulna (forearm), the tibia (above the hock), and around the hock joint itself. If you suspect that your goat has sustained serious damage to its limbs, attempt some basic first aid yourself, mainly attempting to keep the kid immobile so that it does not cause itself more damage, and also remove other inquisitive goats if possible, before calling your vet. Your vet may wish to take X-rays to assess the damage, but if there is little displacement of bone ends your vet may simply immobilize the damaged limb in a plaster cast. With more severe damage, then some form of internal fixation may be attempted, such as pinning or plating, although this is an expensive procedure, and is only likely to be undertaken on valuable or pet goats.

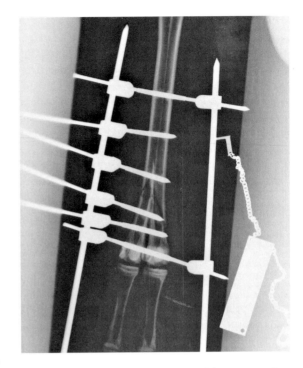

Internal fixation of a fractured limb by means of a plate and screws.

Joint Ill

Cause: Many organisms present in the kidding pens can potentially gain access to the kid's body via its navel, including *Streptococci* and *E. coli*. Prevention has already been discussed in Chapter 6, but to reiterate, it is important that kids are born into a clean environment, that the navel is dipped or sprayed with a strong iodine solution, and that the kid receives adequate colostrum. Organisms entering the body via this route have a tendency to settle in joints, pus then forms in the joints, causing increasing pressure to build up, and the pus itself can erode away the joint capsule.

Signs: Affected kids are very stiff and reluctant to move; they often stand with an arched back if more than one joint is affected. They normally stop feeding, and if suckling, often have difficulty keeping up with the doe, and may become separated if outside. Their joints are visibly enlarged, and are painful if gentle pressure is applied; the kid may be running a temperature, and the navel may also be thickened.

Treatment: Discuss this with your vet; there are a number of antibiotics that may be effective, but to be successful, any treatment regime must be instigated quickly; a prolonged course of treatment is usually necessary. With valuable or pet kids, your vet may decide to adopt a more robust approach, by draining affected joints and injecting antibiotic directly into the joint itself.

Prevention: Ensure that all the preventative measures discussed in Chapter 6 are followed:

- Maintain a clean environment for kidding.
- Dip the navel in strong iodine solution, or similar.
- Ensure that the kid has adequate colostrum, at the right time.
- Recognize affected kids early, and give prompt effective treatment.

Mycoplasma Arthritis

Mycoplasma arthritis occurs in young kids (usually under six months of age) in many countries, but is not currently a problem in the UK, although with increased movements of animals and the opening of international boundaries this may not always be the case. The main organism involved is *Mycoplasma capricolum*, and although outbreaks of disease are normally associated with severe arthritis, other signs, including high temperatures and pneumonia, have also been reported.

Erysipelas Arthritis

This is a bacterial arthritis caused by *Erysipelothrix rhusiopathiae*, a soil-living organism that can be a problem in goats, particularly if the unit also carries pigs, in which the organism can be a real problem (pig muck spread on land can raise the background level of infection). Although Erysipelas arthritis can be part of the juvenile kid joint-ill picture, gaining entry via the navel, it can also occur as a specific entity in older kids, the organism entering the body through natural wounds, or those associated with castration. Affected kids develop hot, swollen and very painful joints, and are often sick, running a high temperature.

White Muscle Disease (Muscular Dystrophy)

Cause: This condition is related to a deficiency in the diet of the pregnant doe of selenium and/or vitamin E. This can be associated with the feeding of poor quality hay or straw, and if soil on the farm is naturally low in selenium, then any homegrown feed constituent is also likely to be deficient. It has also been reported that diets rich in unsaturated fatty acids such as fish/soya oils can cause a relative vitamin E deficiency. The result is that any kids born to a deficient doe may be very low in these essential 'micro-nutrients', and the situation is made worse because milk does not supply an adequate amount. Vitamin E and selenium interact together, and are important for, amongst other things, the normal functioning of muscles.

Signs: Kids can be affected at birth, or up to about four months of age. If kids are born indoors and turned out to grass, it is often the most active ones that are affected first, usually two or three days after turnout, although the problem can also occur in housed kids.

There are essentially two forms, depending on which muscles are affected. The commonest presentation is of a kid that is reluctant to move or is recumbent. Movement appears painful, and affected kids will often cry in pain, although they will normally continue to feed. If they do move, it is with a very stiff-limbed gait. The second form is 'sudden death' due to damage occurring to the cardiac muscle, with resultant heart failure. If found in the early stages of disease, kids are usually dull and depressed, and have a fast heart rate and difficulty breathing as heart failure sets in. Your vet will be able to distinguish this disease from others such as joint ill that may show similar signs.

Treatment: Discuss this with your vet. Supplementing with vitamin E and/or selenium may help, but the muscle damage that has already occurred must be allowed to heal naturally, and this will take time. Nursing is vitally important, to ensure that affected kids are not trodden on, and can escape being taken by predators such as foxes if they are outside.

Prevention: If the problem is identified in kids in their first weeks of life, then it is important that any pregnant dams are given a combined vitamin E and selenium preparation, to boost the reserves of any developing kids before they are born. In older kids, a similar approach would be directed towards the 'cohort group' – that is, those goats of a similar age to the ones affected. Discuss with your vet the best way of doing this; preparations to consider include injectable preparations, drenches or boluses.

Problems Related to Poor Bone Mineralization

Cause: This condition is related to an imbalance of calcium, phosphorus and vitamin D in the diet.

Signs: There are a number of syndromes related to this imbalance that can be recognized, particularly in growing kids. These include (usually in rapidly growing young kids):

- Bent limbs, related to uneven bone development at the growth plate.
- True 'rickets', in which there is painful enlargement of the ends of the bones.
- Bone softening (referred to as 'osteodystrophy'), resulting in weak bones that may fracture.

The signs will be similar in each presentation, including lameness and discomfort, an uneven gait, and stiffness or arching of the back.

Treatment: Correct the calcium/phosphorus imbalance, and some affected kids will respond to nursing and physiotherapy; however, severely affected goats may need to be destroyed on humane grounds.

Prevention: Examine the diet, and ensure that the calcium/phosphorus balance is correct, and allow plenty of exposure to sunlight for vitamin D synthesis. Consider supplementing the cohort group if blood tests show any abnormality.

CHAPTER TWELVE

The Skin and Coat

Unlike problems that may develop with other organs, the skin and coat are readily visible to the observer, and abnormalities can be quickly recognized. The quality of the coat is a good indicator of general health, and a poor coat with broken hairs can be an early sign of debility.

This chapter will concentrate on specific skin problems that may be encountered, but before considering the causes (and there are many), it is important to assess the overall situation before deciding if any treatment or control method is needed, consulting your vet if you are concerned. Consider the following questions:

- Is there hair loss? If so, is it generalized – hair falling out over most of the body – or is it patchy, are bald areas developing?
- Is the goat itchy, or what is more correctly referred to as 'pruritic'? Is it rubbing against the building, gates, fences, or is it nibbling itself? If yes, is it generally itchy, or is it focusing attention on only one part of the body, such as its feet? Remember, all goats enjoy a 'good scratch', but is it being carried out to excess?
- Is more than one goat affected? If several start rubbing and scratching together, it may indicate an infectious problem such as lice.
- Have any goats been recently introduced into the affected group? Could they have

brought in, for example, lice or mange?
- With angora (or other fibre goats), have they recently been shorn, has an outside contractor been involved, could he have inadvertently introduced something?
- Has anything been applied to the goats or their surroundings recently that may be irritant to the coat?
- What part of the goat is affected: ears, feet, head?
- What does the skin look like? Is it dry, crusty, moist, bleeding?
- Can you see anything moving? Many ecto-parasites that live on the surface of the skin are readily identifiable with good eyesight or a hand lens.
- Are any humans in contact affected? Many skin conditions are zoonoses (*see* Chapter 17).
- If in doubt, contact your vet; some conditions can be diagnosed by clinical examination, but your vet may take skin scrapings, plucks of hair or other samples for laboratory examination to identify a cause.

Skin conditions can be conveniently broken down into a number of categories, depending on the cause, although it is important to remember that many skin problems can be multi-factorial (that is, involve a number of different causes).

ECTOPARASITES

This term refers to those parasites found on the outer surface of a host, as opposed to endoparasites (such as worms) that live inside the body. Within this category, disease is associated mainly with lice and mange mites.

Lice

Two types of lice are described, and these are visible to the naked eye. Biting lice have a rounded head end and cause intense irritation, whereas sucking lice have a pointed head end and can suck blood, causing anaemia in addition to surface irritation. Lice infestation (often referred to as 'pediculosis') is often more of a problem in goats already debilitated due to other problems such as worms or an unsuitable diet.

Mange

A number of different mites cause the condition that is collectively referred to as mange; some live on the surface of the skin and are just visible to the naked eye, others burrow into the skin surface and are only visible by using a microscope and examining skin scrapes. Your vet may decide to take a skin scrape in order to arrive at a diagnosis.

The clinical signs and distribution of lesions related to ectoparasitic disease are summarized in the table below.

Other signs not referred to below include:

- Anaemia and debility: sucking lice.
- Aural haematoma formation, presenting as a swollen thickened earflap, and caused by excessive shaking of the head leading to blood vessel rupture under the skin: psoroptic mange.
- Secondary infection resulting in 'pimples' on the skin surface: demodectic mange.

Some of the ectoparasites listed below, particularly the mite causing sarcoptic mange, are zoonotic, and can be picked up by persons handling infected goats.

Treatment: There are many preparations available for treating ectoparasite problems, although the number of licensed products to treat goats is minimal, particularly in the UK. Your veterinary surgeon will be able to advise you on the best course of treatment to use.

Prevention: Although mites can live off the host for short periods of time in the environment, the main source of infection is an infected goat! If you are introducing new goats, make sure they go through a period of quarantine (a

Condition	Cause	Distribution of lesions	Pruritus (itchiness)
Lice	*Damalinea caprae* (biting) *Linognathus stenopsis* (sucking)	Head, neck and back	+ +
Chorioptic mange	*Chorioptes caprae*	Lower limbs and occasionally ventral abdomen or belly	No
Sarcoptic mange	*Sarcoptes scabei*	Often starts on head and ears, then moves to body	+ + +
Psoroptic mange	*Psoroptes cuniculi*	Ears – causes crusting in the ear canal, leading to head shaking	+ +
Demodectic mange	*Demodex caprae*	Multiply in the hair follicles and sebaceous glands in the skin. More common in young goats. Sites affected can be variable, often widespread	No

Clinical signs related to ectoparasitic disease.

A biting louse under high magnification.

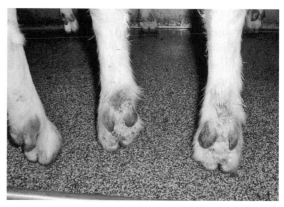

Chorioptic mange tends to affect the lower part of the limbs.

minimum of two weeks) in case they are incubating disease. Treat under veterinary supervision if there is any doubt.

Fly Worry

Biting flies can be a particular problem in housed goats during warm summer weather. They can cause quite severe and widespread lesions over the body, with some goats apparently more susceptible than others. The lesions are usually intensely itchy. Blowflies are much less of a problem than in sheep, but wounds – including those related to disbudding/dehorning, footrot lesions or areas of the body heavily soiled with urine or faeces – can be subject to 'fly strike', resulting in a maggot burden developing. Discuss treatment and control with your veterinary surgeon.

VIRAL DISEASES

Contagious Pustular Dermatitis: 'Orf'

Cause: A parapoxvirus.

Is it infectious? Orf is a very infectious condition affecting both goats and sheep, with the continuing risk of infection spreading between the two species if they are kept together. The scabs drop off into bedding or into the general environment, and can remain infectious in a sheltered environment for two or more years;

however, the virus is destroyed by extremes of weather such as bright sunlight.

Age group affected: Young kids are most often affected, but the author has seen well-established infection in susceptible adults. Outbreaks of disease can develop, particularly in young kids; these can be affected after contact with infected teats when suckling.

Main signs: The lesions begin as pustules, gradually developing into crusty, scabby lesions mainly on the commissures of the lips, gums, nostrils and the lining of the mouth. More unusual sites include the udder (a real problem in milking goats, as lesions can become secondarily infected – *see* Chapter 14), the feet and limbs. When the scabs are knocked off, the exposed underlying tissue is very inflamed and painful. Mouth lesions in young kids can prevent normal feeding behaviour, with debility a potential sequel.

Treatment: If you suspect that your goats may be infected with orf, it is important to make sure that you are well protected before handling them, as humans can readily pick up infection: *wear a pair of gloves*! Although the lesions are fairly characteristic, your vet may decide to send samples into a laboratory to examine scab material for the virus particles. There is unfortunately no treatment available, as this is a primary viral condition, and will almost always resolve naturally. Antibiotic creams and sprays are often used to control secondary infection, and nursing is vitally important, particularly in young kids that may be unable to suckle properly. Watch out for mastitis in the doe if she develops orf lesions on the ends of her teats.

Prevention: If your goats have never previously experienced orf, then measures should be taken to keep it out, by preferably buying any replacement goats only from known orf-free herds wherever possible. A quarantine period is useful to allow new goats possibly incubating disease to develop it safely. Do not be tempted to vaccinate a clean herd, because the vaccine used is based on a live virus, and you run the risk of infection spreading before immunity has developed.

If orf is a problem in one year, then it may be worth considering vaccination in subsequent years, but discuss this approach with your vet. It is recognized that outbreaks of disease on known infected farms are often associated with browsing or grazing infected areas that have a population of potentially abrasive plants such as thistles and brambles, both of which can damage the lining of the mouth, thus allowing entry of the virus.

Human health risk: As already stated, the condition could be readily picked up by anyone handling an infected goat (*see* the illustration on page 157). Take particular care with small children, as a lick from an infected goat may lead to unpleasant lesions on the face of the child.

Goat Pox

True 'goat pox' is caused by a Caprivirus, and does not currently occur in the UK, where it is a notifiable disease (*see* Appendix I). It is confined mainly to Africa (north of the equator), the Middle East, Central Asia and India. It is a highly contagious disease beginning with a high fever, salivation, nasal discharge and conjunctivitis. Skin lesions can develop, and death can occur as a result of secondary infection. In the UK the term 'goat pox' is often used to describe a Staphylococcal skin infection (*see* below): the two must not be confused!

Scrapie

Although not a true viral disease, scrapie is considered under this heading. It can cause pruritus and self-mutilation by excessive chewing and nibbling, although the skin form is not as well pronounced as in sheep. Scrapie is considered in Chapter 10.

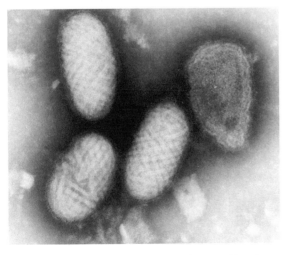

The virus causing orf, seen on high magnification using an electron microscope.

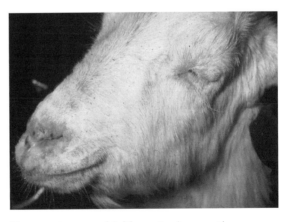

True goat pox – a highly contagious exotic viral disease of goats (not to be confused with Staphylococcal infection, often referred to colloquially as 'goat pox').

BACTERIAL DISEASES

Staphylococcal Infection

Cause: *Staphylococcus aureus* is considered a normal inhabitant on the skin of many goats, but with the capability of causing secondary infection of wounds; if it gains access to the hair follicles it will cause multiple skin abscesses. Demodectic mange (*see* earlier) may act as a precursor to infection.

Clinical signs: The udder is a common site for such abscesses to develop, and the term 'udder impetigo' has been used to describe the distribution of lesions. They can develop either in the area between udder and upper limb where the two surfaces rub, or at the teat end as a sequel to damage in that area. This may be the result of a faulty milking machine, to over-milking, or to the rough removal of the claw piece when milking has finished; it can also sometimes follow damage resulting from a vice referred to as teat biting, in which individual goats appear to derive satisfaction by biting and chewing the teats of other goats. This vice can be a real problem in commercial herds, and culprits are best identified and culled (*see* Chapter 14). Spread from goat to goat can potentially occur as udders are prepared for milking, and there is the added risk that infection may gain entry through the teat end causing mastitis.

Abscesses of varying sizes, from a few millimetres to 60mm, have been described in many other sites on the body, and these often feel roughened as they burst and cause a scab to develop. Goat owners often refer to this as 'goat pox', but this is a misleading term (*see* Goat Pox, p. 135).

Treatment: Any treatment is often disappointing, although your vet may prescribe washes and shampoos that may help. A prolonged course of antibiotic should be avoided, as it will upset the normal balance of micro-organisms in the gut.

FUNGAL DISEASES

Ringworm

Cause: A fungal infection affecting a wide variety of animals including pets, caused in ruminants usually by *Trichophyton verrucosum*.

Is it infectious? Ringworm is very infectious, either by direct contact between infected goats, or by indirect contact with spores on gates, fences, door frames and so on, where other infected animals (particularly cattle) may have been rubbing. It is not a common problem, however, although a number of goats on an 'infected' unit can show lesions.

Age group affected: Infection can occur at almost any age.

Main signs: The typical ringworm circular lesion with hair loss and crusting can be seen almost anywhere on the body. In the author's experience, however, lesions in goats can be more pruritic and aggressive than in other species, with a moist 'eczema'-like appearance. Lesions are also more commonly encountered on the head, ears and neck in goats. As lesions heal, they often leave a circular 'bald' patch. Although lesions are fairly characteristic, your vet may take samples to confirm the diagnosis.

Treatment: Discuss this with your vet; there are a number of sprays and washes available.

Human health risk: This is yet another skin condition affecting goats that can be picked up by humans who come in contact with them. *Wear gloves* therefore when handling them – and remember that any bare skin coming into contact with ringworm lesions can become infected, so beware handling goats when you are wearing shorts and a tee shirt or skimpy clothing!

Prevention: This is aimed mainly at keeping infection out, based on quarantining all incoming animals. Also avoid housing goats in buildings that may have previously housed cattle, in which ringworm is more widely reported.

Ringworm lesions on the head of this goat; note the moist eczematous appearance compared to the more normal circular dry lesion.

MISCELLANEOUS SKIN CONDITIONS

Pygmy Goat Syndrome

This disease is confined to pygmy goats (as the name suggests), and is a fairly ill-defined skin condition, tending to develop mainly in goats in their first year of life. Lesions occur on the ears, nose, axillae (the area between the elbow and chest) and in the area around the groin and tail. Hair loss and skin crusting is evident with or without pruritus. It is thought to be associated with a disorder of keratin production in the skin surface. It appears to be hereditary, and often occurs in families.

Your vet will confirm this condition, mainly by eliminating other causes of skin disease, but any young pygmy goat with a poorly responding skin condition is a likely candidate for this ailment. No specific treatment is effective, although some baths and washes may help. It is possible that zinc deficiency may make some cases worse (*see* later).

Labial Dermatitis

Artificially reared kids, particularly those drinking from a bucket, may develop a moist eczema with hair loss and scab formation around the muzzle on the lips. It appears to be related to an accumulation of fat from the milk around the muzzle itself, and secondary infection is a possible sequel.

Golden Guernsey Goat Syndrome ('Sticky Kid')

This syndrome is another hereditary breed problem: affected kids are born with sticky, greasy matted coats that remain abnormal for life.

Note the thickened scurfy skin typically seen in cases of the 'Pygmy Goat Syndrome'.

Zinc Deficiency

Zinc is a trace element involved in many metabolic pathways in the goat's body, including keratin production. A deficiency of zinc in the diet can lead to a thickening of the skin, particularly over the hind end, and the head and neck, with associated hair loss. It may be associated with the pygmy goat syndrome (above). Discuss possible supplementation with your vet, but nutritional supplementation with, for example, zinc sulphate may be advised.

Urine Scald

Male goats will frequently urinate on themselves, particularly during the breeding season. This will result in hair staining, particularly down the back of the front legs and over the face, and scalding of the skin can occasionally occur. Urine scalding is also a problem in any goat that is recumbent for any period of time – they should be regularly moved and kept clean.

SKIN SWELLINGS

Tumours

Although not common, these can occasionally develop. Tumours of the lymphoid cells (referred to as 'lymphosarcoma') present as enlarged lymph nodes, many of which are clearly visible under the surface of the skin; lymph glands under the jaw are particularly affected (*see* Chapter 9). Other occasional tumours include papillomas (warts), with the head and udder being sites most often affected.

Caseous Lymphadenitis

See Chapter 9.

Injection Site Reactions

Goats will often develop swellings at the site of an injection (particularly with certain vaccines, for example Clostridial). Although these can occur if the vaccination technique was unhygienic, some goats suffer a local reaction to the oily solution in which the vaccine

An injection site abscess – these can be very large and unsightly.

is suspended. These may persist for many months, and can be unsightly.

Neck Swelling

Some goat kids may develop a swelling on the underside of their neck, in the first few weeks after birth. This appears to be a normal developmental 'spurt' in the growth of the thymus (lymphoid tissue), which usually regresses by about six months of age.

Wattle Cysts

These are inherited, occurring mainly in Anglo Nubian and British Alpine goats, in which a swelling develops at the base of one or both wattles under the chin. They can grow to be quite large and unsightly (several centimetres), although they are relatively harmless. They may be safely removed if unsightly; discuss this with your vet.

Swelling Disease of Angoras

A condition reported worldwide in Angora goats (mainly young kids), and thought to be linked to vitamin E deficiency, with the Angora goat possibly having a higher requirement for

Ticks – capable of transmitting a number of diseases between goats (and other farm/wild animals).

this vitamin than other breeds. The typical signs are swellings under the skin of one or more limbs, and any/or all of the following areas: under the jaw, lower neck area, lower part of the thorax and the abdomen. The swellings are soft and fluctuant, and may vary in size from day to day before resolving spontaneously. Affected kids are often anaemic, showing marked pallor of the visible mucous membranes.

Ticks

Goats out on rough grazing may occasionally pick up a tick that attaches itself to the skin, sucks blood for a week or so, and then drops off again. They can be quite large and alarming when seen for the first time, but you should resist the temptation to pull them off: the head and mouthparts can be left behind, and can cause an unpleasant local reaction. They are of minor significance, although if the tick bite becomes infected, it can lead to spread of infection via the bloodstream, leading to abscess formation in many sites, including joints and the spinal column (tick pyaemia). Ticks can also act as vectors of certain diseases (as they suck blood from different hosts during their lifetime); one that is occasionally encountered in goats, and in which fever is the main presenting sign, is referred to as tick-borne fever.

CHAPTER THIRTEEN

The Eyes and Ears

THE EYES

The eyes of a healthy goat should be bright and clear. Eye problems may be indicated by any of the following:

- Excessive discharge or weeping: is the discharge clear, or is there pus present?
- Excessive blinking.
- Cloudiness of the eye.
- Sunken eyes, often a sign that the goat is unwell, and a feature of excessive weight loss or dehydration.
- Flickering of the eyes from side to side (nystagmus), a feature of listeriosis (*see* Chapter 10).
- Lack of a 'blink response' – a normal goat will flinch and blink its eyelids when a hand is quickly passed in front of its eyes. A sick goat will have a poor response.
- Apparent blindness – the goat may bump into objects in its path.

Entropion

Although a relatively common problem in newborn lambs, it appears to be a lesser problem in young kids. In entropion, one or both eyelids are rolled inwards so that the normal edge of the lid is not visible. This results in the lashes and hairs on the skin surface rubbing over the front of the eye, causing intense irritation and abrasive damage to the eyeball if not spotted early. Affected kids will rapidly become blind and be unable to find the udder and feed if suckling.

Although a minor in-turning of the eye can be corrected by gently rolling the lid back, it tends to recur. Your veterinary surgeon will carry out a minor surgical procedure to correct the problem. Entropion undoubtedly has a genetic element, and if a number of cases are identified in any season, it may be worth checking to see if they were all the offspring of the same buck, or if the dams are related.

Infectious Kerato-Conjunctivitis

Cause: A number of micro-organisms are capable of causing this condition, including *Mycoplasma*, *Listeria* and *Moraxella*. Outbreaks can occur and spread if either a number of goats are kept close together (with limited trough space, for example), or when they are covered in flies that are attracted by runny eyes, and thus readily transmit infection from goat to goat.

Age group affected: This is quite variable, but true infectious kerato-conjunctivitis tends to occur in older goats. Young growing kids and goatlings can show a mild conjunctivitis as part of an upper respiratory tract infection involving *Mycoplasma*.

Main signs: One or both eyes may be affected, and typically there is excessive blinking, and discharge from the eyes with tear stain deposits on the side of the face. The area around the cornea (the sclera or 'white of the eye') is reddened, and the condition is often referred to as 'pink eye' because of this appearance. The eye itself often appears cloudy as if covered by a milky film, and with time, small blood vessels can be seen growing in from the sclera towards the centre of the eye. Although this is actually the natural healing process of severe eye damage, it will often result in total blindness. If the eye recovers, then sight may return, but there may be scarring of the cornea.

Treatment: Discuss this with your vet; the actual treatment used depends on the severity of the problem, with mild cases responding spontaneously without treatment. Your vet may prescribe antibiotic ointment or powder to place over the surface of the eyeball; more severely affected cases will benefit from injectable antibiotic either intravenously or intramuscularly, or into the conjunctiva around the eye, when the antibiotic will gradually leak across the eye from the residual needle hole.

Prevention: This is not easy, as many apparently healthy goats may well carry the organisms involved. If cases develop, it may be advisable to try and minimize risk factors for the spread of infection. Try to control flies that act as vectors for the spread of infection, reduce the number of goats kept together to minimize spread, and try to ensure that there is sufficient space for goats to feed together without having to push their heads together to get at feed (the tips of the ears are thought to be a good way of spreading infection between goats in close contact).

Foreign Bodies in the Eye, and Trauma to the Eye

Seeds, wood shavings or sawdust may get into the eyes, and may actually adhere to the cornea, causing intense discomfort with constant blinking, heavy tear staining and a tendency for the goat to rub its eye vigorously. Occasionally, foreign bodies may become trapped behind the third eyelid in the corner of the eye. Unless the foreign body can be easily removed without causing further damage or distress, contact your vet, who can administer anaesthetic drops into the eye, to deaden the pain, although it may also be necessary to sedate or even anaesthetize the goat to properly restrain it and prevent it from struggling. Dyes such as fluoresceine are often placed in the eye to check for damage to the cornea.

Due to goats' inquisitive nature, and ready ability to clamber over, under, on top of or through obstacles and barriers, there is a constant risk of damage to the eye by protruding branches, wire and so on. Damage may be serious, and your vet may need to suture the lids together to prevent further injury and allow the eye to heal. If the damage is severe, enucleation (removal) of the damaged eye may be necessary – this is a surgical procedure that can produce a dramatic response, if the goat is in great pain.

Blindness

The way a goat behaves should alert you to the fact that it has gone blind, although a goat that goes blind suddenly will be easier to spot than one that goes blind over a period of days or even weeks. The goat that suddenly goes blind will stand still as if frightened to move, and if it does move it will bump into obstacles in its path. When a goat goes blind gradually, it can appear quite normal in its own environment, as it has 'learned' to find its way around. When placed in an unfamiliar environment, however, it will behave like a goat that has suddenly gone blind, bumping into obstacles in its path.

When a goat goes blind, there are a number of reasons why, and not all these will be obvious. If the cornea has been damaged due to trauma or infection, or if the goat develops a cataract, this damage can be seen. But if the damage is to the retina, or if it originates in the brain, then a detailed examination of the eye by a veterinary surgeon to determine a cause is necessary.

THE EARS

The carriage of the ears when held erect gives a picture of alertness; conversely if the ears are drooping then the goat may be feeling unwell (except in the case of the droopy-eared Anglo Nubian), and these subtle changes can be picked up by an experienced owner.

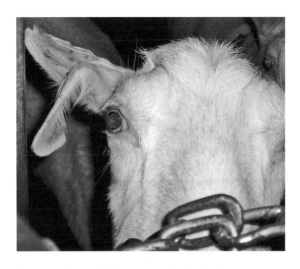

Ears can be readily torn, particularly if an eartag gets caught in, for example, a fence.

If only one ear droops, and the other ear is held normally, it may indicate a CNS problem such as Listeriosis (*see* Chapter 10). If there is excessive head shaking, this may indicate ear irritation by ear mites, or a foreign body in the ear canal.

Ear Mites

Caused by the ear mange mite *Psoroptes cuniculi* in the UK. Many goats may carry the parasite with few clinical signs, but in susceptible goats or in those with a heavy infestation, clinical signs may develop. These may vary from an occasional shake of the head, to violent shaking, and rubbing the side of the head against solid objects. Dried scab material may be seen in the ear canal, mixed with blood if self-inflicted trauma has occurred. Kids will pick up infection from their dams, and may show clinical signs as early as three to four weeks of age.

Possible complicating problems include an aural haematoma in which the violent head

shaking results in rupture of small blood vessels under the skin of the ear flap, which swells as a result. As this swelling subsides, the ear becomes contracted and twisted (a so-called 'cauliflower ear'). Very occasionally the ear canal can become infected, leading to damage to the ear drum, which in turn may predispose to middle ear infection, in which the head is held permanently twisted to one side or the other. Your veterinary surgeon will advise the use of an anti-parasitic solution to place in the ears.

Foreign Bodies

Although rare, it is possible for goats to get grass seeds or barley awns in the ear canal, particularly if they are burrowing their heads into hayracks or hedgerows. This will result in violent head shaking, and your vet may have to administer a sedative to retrieve the foreign body using forceps, possibly aided by an auriscope (an instrument used to look into the ear canal).

Trauma

Ears can be torn if caught on barbed wire, projecting metalwork or thorns, and the damaged ear may bleed profusely, covering the head end of the goat with blood (particularly if they shake it). Don't panic, however, as the damage is usually not as bad as it seems, and the bleeding will normally stop spontaneously; if you are concerned, contact your veterinary surgeon.

New legislation has come into place across the EU regarding eartag insertion. Goats appear to tolerate eartags less well than sheep, and seem to suffer more discomfort when the tag is inserted, with a higher proportion developing reactions at the site of tagging. Trials are currently being undertaken to assess different types of tag at different sites to overcome this.

CHAPTER FOURTEEN

The Udder

Although of particular importance in dairy goats, udder problems can develop in any type and breed of goat regardless of its potential milk yield. Mastitis (inflammation of the udder) is the most serious problem affecting the udder, although it can vary from a serious condition in which there is a rapid onset of severe toxaemia and death, to mild cases where udder change is minimal or not apparent at all.

STRUCTURE OF THE UDDER

The udder consists of two halves (as opposed to four quarters in a cow), with each half consisting of a mass of glandular tissue responsible for milk secretion. The milk is actually produced by cells lining the alveoli – small sac-like structures deep within the mammary gland. Surrounding the alveoli are muscle cells, which contract, squeezing milk from the alveoli into the ducts. The milk collects in the duct system at the base of the gland, and is then removed from the udder via the teat (either by a kid sucking or milking by hand or machine), with the teat end protected by a teat sphincter preventing leakage of milk or entry of bacteria.

DISEASES OF THE UDDER

Mastitis

As already stated, mastitis is defined as an inflammation of the udder or mammary gland,

and is further classified as per-acute, acute, chronic (all classed as 'clinical mastitis') and sub-clinical, depending on its severity. Compared to other milk-producing animals such as cows, however, clinical mastitis is comparatively rare, but sub-clinical mastitis is very common.

Clinical signs:

- Pyrexia and obvious illness (per-acute/ acute).
- Change in colour of the udder skin surface, to purple/black (per-acute/acute).
- Heat/pain – the udder feels warm to the touch, and is obviously painful (per-acute/ acute).
- Firmness or fibrosis of the individual halves (per-acute, acute or chronic).
- Disparity in size of the two halves (swollen) – per-acute/acute, (shrunken) – chronic.
- Change in milk appearance, with clots or becoming watery in consistency (per-acute, acute or chronic).
- Reduced milk yield – (per-acute, acute or chronic).
- No obvious change to udder or milk secretion – (sub-clinical).

If any of the signs described above are present, then you should suspect mastitis. Assess the goat's general demeanour: does it look ill, is it feeding, is it running a temperature? Goats with per-acute or chronic mastitis will often appear lame, or walk awkwardly due to pain and discomfort. Express milk from the teat:

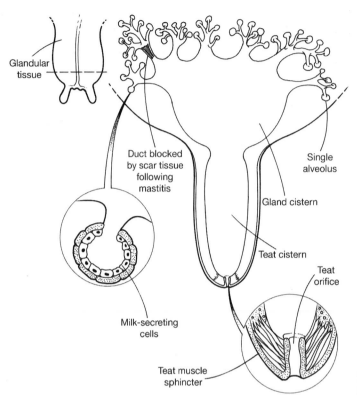

Glandular tissue

Duct blocked by scar tissue following mastitis

Single alveolus

Gland cistern

Teat cistern

Teat orifice

Milk-secreting cells

Teat muscle sphincter

The normal structure of the udder and teat.

does it look abnormal, does it contain clots or blood? If in doubt, contact your veterinary surgeon for help and advice.

Laboratory Tests

Milk samples from an affected udder half can be collected into a sterile container; ensure that the teat end is clean and dry – swabbing with alcohol or surgical spirit will help, before expressing the milk from the teat at a 45° angle into the container. Laboratory examination of a milk sample will give an indication of the bacterium responsible, which in turn will provide clues as to the possible source. Mastitis-producing bacteria are referred to as contagious (mainly spread from udder to udder during the milking process) such as *Staphylococcus aureus*, or environmental (picked up from a dirty environment) such as *E. coli*.

One other laboratory test employed to measure whether or not a goat has mastitis, is the cell count – literally measuring the number

of cells/ml of milk, and this level will rise in response to mastitis as 'pus'. The significance of cell counts in goats is difficult to determine in comparison to cows, however, where there are clearly defined numerical limits. In the bovine udder, a cell count of <200,000 cells/ml is taken as an indication that the udder is uninfected, with counts running up to several million cells/ml in clinical mastitis. The 'normal' background cell count in the healthy goat udder is maybe as high as 1,000,000 cells/ml, and this figure will rise dramatically in response to infection. There are devices that can crudely measure cell counts in milk that are widely used in dairy cow herds to identify infected cows; they are known as 'California mastitis tests' or CMTs, but do not be tempted to use them in goat herds to identify infected goats, as many will fail, yet will not be infected!

Cause: Mastitis will develop when bacteria around the teat end gain access to the udder tissue via the teat canal. This can occur if

A milk sample sent for laboratory examination; note the clots of pus (left) compared to the normal milk (right).

there is damage to the teat end, or if there is a high level of bacteria around the teat end when the teat orifice is open (for example, after milking).

Teat-end damage can occur as a result of a poorly maintained milking machine resulting in teat-end vacuum fluctuation, defective pulsation units, overmilking (leaving the unit on for too long), or rough removal of the unit at the end of milking. It can also occur as a result of

trauma: getting teats caught on jagged edges or wire, or as a result of the vice described as teat biting. Teat-end damage will not only damage the barrier preventing bacteria from getting into the udder, but can also result in scab formation around the teat end with local bacterial multiplication occurring.

High levels of mastitis bacteria around the teat end can occur if:

- There is localized skin infection of the udder, for example Staphylococcal infection of the udder skin, wounds resulting from machine damage, teat biting or fly damage.
- The milking cluster becomes contaminated following the milking of a previously infected goat.
- There is environmental contamination of the teat ends with faecal material.
- An unhygienic milking routine is used, involving dirty hands, poorly cleaned milking clusters and so on.

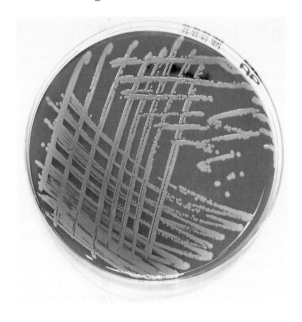

A culture plate showing colonies of Staphylococcus aureus.

This is the crude test for high cell count in milk, referred to as the California Mastitis Test (CMT); it has limitations in goats and should be used with care.

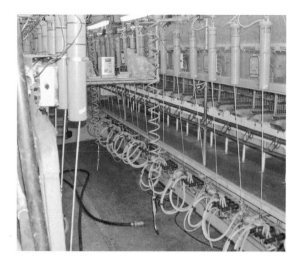

Milking clusters can be contaminated with mastitis bacteria during milking.

Treatment: Discuss this with your vet. The approach depends on the severity of infection, the numbers infected and the organisms involved. Mild cases may respond to local treatment using an intramammary preparation of antibiotic infused into the affected half of the udder, after the gland has been 'stripped out' – that is, all milk/pus has been milked out through the teat orifice. It normally requires a minimum of three tubes infused at twelve-/twenty-four-hour intervals.

More severe cases will undoubtedly require a more aggressive treatment, including injectable antibiotics, and supportive therapy such as anti-inflammatory preparations; very sick animals will benefit from intravenous or oral fluid therapy.

The mortality rate from acute gangrenous mastitis can be quite high, and many surviving goats may need to be culled on humane grounds if the udder damage is severe: parts of the udder can literally die and begin sloughing away. Take precautions to prevent fly worry and maggot strike if this occurs during hot weather.

Prevention: Many commercial large and small units have low levels of mastitis, yet undertake only minimal control measures. If, however, problems develop, then the following list gives an indication of the available control measures that may be adopted:

- Keep goats in a clean environment, to prevent faecal contamination of the teat end.
- If goat teats are clean prior to milking, avoid washing the teats. They will need to be washed if they are dirty, however, using a proprietary teat wash preparation; ensure that only the teats are washed, not the whole udder. The teats must then be dried thoroughly with a paper towel (one per goat) before the cluster is applied or the goat is milked by hand.
- Wear disposable gloves during milking, and change these if they become torn or dirty.
- Palpate the udder before milking to check for swelling, and ideally strip a small quantity of milk from the teat end into a 'strip cup' to check for mastitis.
- It may be advisable, particularly if you have a problem with mastitis, to use a proprietary teat dip or spray after milking is complete, to kill any bacteria around the teat end (which will still be open); if it contains, for example, lanolin, it will also help to maintain the quality of the teat skin.
- Treat cases promptly when recognized, marking, then segregating them, ensuring that they are milked last.
- Cull chronic or incurable cases; they will only act as a reservoir of infection.
- Consider the use of 'Dry Goat Therapy', in which a long-acting intramammary antibiotic preparation is inserted into the udder as the goat is dried off at the end of lactation. This has the dual effect of treating existing infections, and preventing new ones being picked up during the dry period.
- Ensure that records are kept of mastitis cases, in order that repeat cases can be identified and culled if necessary.
- Maintain the milking machine so that it runs correctly, and keep it clean between milkings.

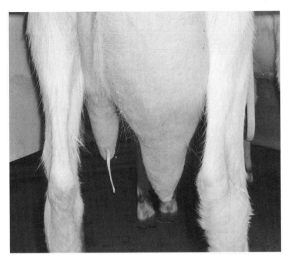

The left half of this udder has become shrunken and firm due to CAE viral damage.

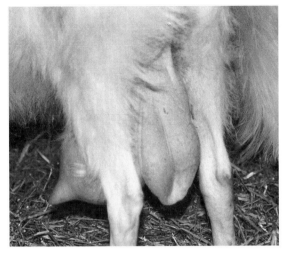

Some male goats can develop udder tissue, and even begin lactating.

CAE Indurative Mastitis

A relatively common feature of CAE infection is a form of mastitis in which the milk remains unaffected, but the udder tissue gradually shrinks and milk production is reduced (*see* Chapter 11).

Mycoplasma Mastitis

Although not currently a problem in the UK, mycoplasma mastitis is a major problem in many parts of the world, and potentially could be imported into the UK, although there is a national monitoring scheme in place. Although a number of mycoplasma organisms are capable of causing mastitis worldwide, it is *M. agalactiae*, the cause of contagious agalactia, that is economically the most significant mycoplasma mastitis, being confined mainly to the Mediterranean area. Infection can spread rapidly within a group of goats, and other signs including arthritis and conjunctivitis may be seen. Milk changes vary from none, to watery, to very thick and purulent, the udder becomes firm, and the milk supply dries up.

Udder Oedema

Oedema is the accumulation of fluid in body tissue, and this can occasionally occur in the tissue of the udder in the last few days before kidding as a normal feature. It can become excessive, however, and cause gross swelling with pain and discomfort, although the condition will resolve naturally when kidding takes place and the goat is milked out. Discuss the problem with your vet if the doe is in obvious discomfort, but unless advised to do so, resist the temptation to milk the udder before kidding.

Udder Development Unrelated to Pregnancy

Maiden goats that have never bred may show udder development, and may even produce milk. Unless you intend to milk the goat daily to encourage a full lactation to develop, do nothing: removal of the milk will result in more being produced, and increase the risk of mastitis developing. If the goat is obviously distressed or uncomfortable, consult your vet. This phenomenon is mainly linked to the offspring of particularly high-yielding goats.

Male goats may also occasionally show udder development, and reducing the dietary input during the main risk period, which tends to be the summer months, can control the condition. Severe mastitis is a potential problem, and the condition should be monitored – fertility is not usually affected.

Trauma

The udder is quite prone to damage, at risk of being butted, the teats being trodden on, getting caught in gates or fences and suchlike, any of which can result in pain and discomfort; torn or cut teats in particular may bleed profusely and need suturing.

Tumours

Although not common, tumours may develop in the udder tissue, particularly of elderly females.

Milk Taints

See Chapter 16.

UDDER AND TEAT SKIN PROBLEMS

Teat Biting

This is a vice that can sometimes develop in individual goats, and can cause quite severe damage to teats, resulting in mastitis. Offenders must be identified and removed.

Staphylococcal Infection (Udder Impetigo)

Small pimples can develop on the skin of the udder and teats, either solely as an udder problem, or as part of a more generalized Staphylococcal infection (*see* Chapter 12). Pimples can coalesce to give areas of moist dermatitis, particularly between the udder and the inside of the leg. These infected lesions can act as a source of *Staphylococcus*, resulting in mastitis.

Contagious Pustular Dermatitis (Orf)

See Chapter 12.

Foot and Mouth Disease (FMD)

Lesions of FMD can develop on the skin of the teats, and in the FMD epidemic in the UK in

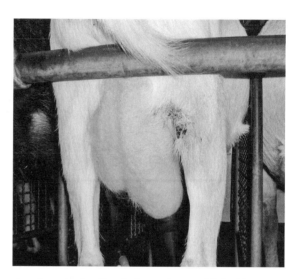

Another site for Staphylococci to multiply (the source for mastitis) is in the area between the inside of the limb and udder, resulting in a moist dermatitis.

2001, teat lesions were the main clinical signs described in one milking herd, with little evidence of the more typical foot and mouth lesions. Early symptoms described included a depression of milk yield, pyrexia and inappetance. Small, often blood-filled blisters developed on the teats, and these quickly burst and coalesced so that the teats appeared red and very sore. Mastitis then followed, with a number of females showing typical signs with milk clots.

This disease is notifiable, and any suspicion of FMD must be reported to the relevant authority (*see* Appendix I).

Supernumerary and Abnormal Teats

All kids should be checked for teat abnormalities at birth or when disbudded. Supernumerary teats are additional teats to the normal two, and if uncomplicated are usually removed when the goat is disbudded to prevent possible problems developing in later life. If teats are split or abnormal, these are best left alone, but ask your vet if you are in doubt. Remember that they are inherited defects.

CHAPTER FIFTEEN

The Urinary System

Anatomically the urinary system consists of the two kidneys, the ureters that carry urine produced in the kidneys to the bladder, the bladder, and the urethra along which the urine flows from the bladder during urination. There are few problems that affect the urinary system of the female goat, most problems occurring in males, and related specifically to the anatomical design of the male urinary tract as compared to the female.

Nature has designed male ruminants such as the goat to have a long and narrow urethra that passes from the bladder through the penis, ending in a particularly narrow extension referred to as the urethral process. In castrated males this urethral process is underdeveloped when compared to an entire male, and is a potential site of obstruction should calculi ('stones') develop in the bladder. Fine, sand-like calculi develop quite commonly in the urinary system, but under most circumstances are readily flushed out by the flow of urine, particularly in females. In males, however, and particularly in castrated males, there are three potential sites for small stones to become trapped, and this results in the development of a condition referred to as 'urolithiasis'.

DISEASES OF THE URINARY SYSTEM

Urolithiasis

As already described, this condition is associated with small stones becoming trapped in the narrow urethra of the male, blocking the normal flow of urine.

Animals affected: The condition is most commonly encountered in castrated male kids, and can be seen in kids that are fattened commercially for meat production, and in pet kids. It can occasionally occur in entire bucks, but is rarely, if ever, a problem in females.

Cause: There are many potential causes and predisposing factors to this complex condition, and it is rarely possible to identify any one specific underlying cause. A dietary excess of phosphorus and/or magnesium is commonly identified, and this may be related to a poorly formulated diet, the feeding of an incorrect mix (for example, a diet with high magnesium meant for high-yielding dairy cows at grass) or to the provision of an *ad lib* (free access) mineral mix containing high levels of magnesium/phosphorus, often in a very palatable form, thus encouraging high intake. Another possible reason is low water intake, thus resulting in increased urinary concentration and a greater risk of crystal formation in urine occurring, leading to stone formation. This may be the result of too little water being provided, or the water being contaminated with, for example, faecal material, making it unpalatable. It has also been suggested that goats on a low fibre diet may be more susceptible, as saliva production is not stimulated as much (and saliva contains a high level of phosphorus), thus encouraging phosphorus build-up in the body. In sheep, certain breeds have been found to be more susceptible; however, this has not been

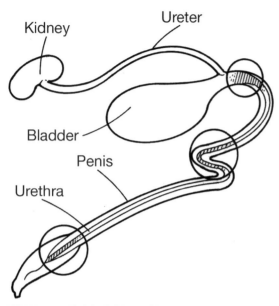

Kidney

Ureter

Bladder

Penis

Urethra

Vermiform (urethral) appendage. Potential sites (circled) for stones to become trapped in the male urinary system.

shown to apply to goat breeds. A particularly unusual type of calculus can develop that has an 'oxalate' chemical make-up, and these may be seen if goats are fed oxalate-containing plants such as sugar-beet tops – sugar-beet pulp itself is quite safe to feed.

Common signs: The presence of stones in the urinary system goes unnoticed until one or more are washed into the urethra by the flow of urine; they can then become trapped at the sites referred to above, thus gradually or abruptly interrupting the flow of urine. As the pressure in the bladder begins to build, the goat will show signs of increasing discomfort. Early signs include kicking the abdomen with the back feet, and a sideways glance at the abdomen. Goats will then lie down, but get straight back up again repeatedly, bleat loudly and begin to strain. This straining is often mistaken for constipation, so look around for freshly voided faeces to rule this out.

The goat becomes increasingly distressed as pressure builds up; its breathing becomes fast, shallow and laboured. Eventually the pressure will lead to rupture of the urethra, either near its tip so that urine escapes to the underside of the belly which will swell noticeably around the penis, or the bladder will rupture and urine will escape into the abdominal cavity. This will appear to bring relief, but the goat will now rapidly deteriorate and die, and treatment, even if attempted at this stage, is often disappointing. It follows, therefore, that the earlier in the course of the problem the condition is recognized, the greater the chance of success. Look at other male goats in the group:

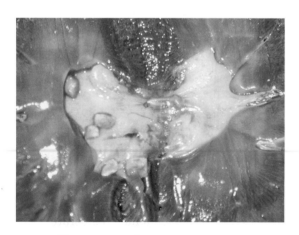

A large stone in the kidney pelvis will often result in no obvious signs of discomfort.

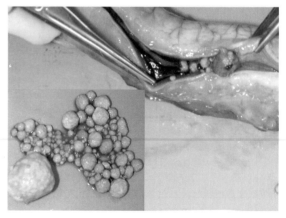

Small stones have passed from the bladder where they have become trapped on their way through the urethra, causing a complete blockage. Inset are the stones found in the bladder at post mortem.

Salt licks can be placed around goat accommodation, and will encourage goats to drink more water.

you may well notice small crystals developing on the hairs around the prepuce, resembling small 'beads' on the hair.

Treatment: This is a genuine emergency, and if you suspect that your goat is affected, seek urgent veterinary attention. There are three possible scenarios that your vet may be faced with: in the first, if there is a dribble of urine escaping, it suggests that the blockage may not be complete, and the likely approach to adopt is a combination of massaging the urethra, the use of injectable products that cause relaxation of muscle (so-called spasmolytic drugs), and the possible surgical removal of the urethral process. If the blockage is complete and the goat is very distressed, surgery is the only alternative, depending on the site of obstruction: either the urethral process is removed, or surgery may be carried out to expose the urethra under the tail.

The third scenario is when the bladder or urethra has already ruptured; in these cases veterinary intervention is almost always disappointing, and your vet may well advise slaughter on humane grounds. Your vet may take blood samples to assess the long-term prognosis, as the kidneys are damaged by back pressure, their ability to function correctly is impaired, and waste products such as urea and creatinine build up in the blood and these can be measured and compared against the reference range for a healthy goat. The presence of one affected goat is, however, an indication that circumstances are such that other cases may develop, and attention should be directed towards preventing fresh cases.

Prevention: Prevention is better than cure! Preventative measures are directed towards management factors as follows:

- Ensure adequate water intake at all times, changing the water twice daily if necessary to ensure it remains clean. Provide warm water on cold days. Ensure that all kids can reach the trough, or can use automatic drinkers if these are provided. Also ensure that there are sufficient water containers to allow all goats to drink together if necessary – and keep an eye on timid goats, making sure that they are not bullied away from water sources.
- Feed good, palatable fodder such as hay or pea straw.
- Feed dried grass products such as lucerne, or a coarse mix to ensure a varied diet.
- Feed a well-balanced diet with a 2:1 calcium: phosphorus ratio.
- Increase the salt content of the ration to encourage drinking, or have a number of salt licks strategically placed.
- Discuss these approaches with your vet.

Cystitis and Pyelonephritis

Bladder and kidney infection can occasionally be encountered in does in the period of time after they have kidded, particularly if they had a difficult kidding, produced dead kids, or developed a prolapse of either cervix or uterus. Cystitis will result in the doe straining, and merely dribbling urine, often with small flecks of blood; they are often very vocal when affected. If pyelonephritis develops, the affected doe is often very ill, will run a high temperature and stop feeding. If you suspect either condition, consult your vet, as antibiotics are required.

CHAPTER SIXTEEN

Has It Been Poisoned?

A veterinary surgeon will often be consulted over a sick animal, accompanied by the phrase 'I think it may have been poisoned'. Although poisoning is a common worry when any animal becomes ill or dies suddenly, in reality it is quite a rare occurrence. It is recognized, however, that goats can consume either toxic plants or chemicals in their environment, and this chapter will describe what may cause poisoning, how to recognize and treat it, but more importantly, how to prevent it from occurring.

Goats are naturally inquisitive, and will investigate and explore new and interesting objects by nibbling, licking or even consuming them, and this particular behaviour can occasionally get them into trouble. Compared to other ruminants, however, they do appear to be able to consume small quantities of potentially quite harmful plants without suffering any apparent ill effects; but as the amount consumed increases, however, a critical point will be reached when problems can develop.

A goat's natural instinct is to browse, to consume a variety of different plants, branches and hedgerow material all in relatively small quantities. If, however, you physically restrict the goats to one area by fencing or tethering them, then this natural browsing behaviour will be prevented, and the risk of consuming a potentially harmful amount of a potentially hazardous plant within easy reach may well increase.

Garden centres now sell many exotic shrubs and plants, and the potential toxicity of many of these is unknown if a goat should gain access to your garden. One particular example is the plant Pieris, which is particularly toxic to goats.

DIFFERENT POISONS AND THEIR EFFECT

- Some poisons (plant or chemical) are highly toxic and will rapidly kill a goat if consumed in a sufficient quantity; classic examples include yew or lead.
- Some poisons rely on a gradual build-up of the toxic principal in the goat's body before signs of illness develop, and copper is a good example. This is often referred to as delayed poisoning.
- Other toxic agents may cause a particular clinical problem such as diarrhoea, a urological problem, or even vomiting, which is very unusual in the goat (see table on p. 153).
- Yet another way a poison may work is to damage a particular body organ or system; although unusual in the goat, acorn poisoning is a good example since the toxic principal causes severe kidney damage.
- Occasionally secondary problems may develop, such as inhalation pneumonia in goats affected by rhododendron poisoning that have inhaled rumen content.
- Milk taint – many plants may cause the goat no harm even when consumed in

large quantities, but have the potential to taint the milk with an unusual flavour or odour.

Some Specific Plant Poisons

Yew: Contains a very powerful toxin that can kill a goat rapidly even after eating only a small quantity. Although goats will not normally consume the branches of this tree, cases can occur if it suddenly appears in their environment, when natural curiosity may lead to branches being nibbled. Poisoning can follow when, for example, the branches of a tree are blown down into a goat paddock after a storm, or when branches have been cut from trees and thrown over a hedge or wall (remember many yew trees are often found in church yards, so be vigilant if you are grazing goats nearby).

Rhododendron/Pieris: Both these plants belong to the same family. Although many rhododendron bushes are found growing wild around the UK, poisoning incidents occur most commonly from cultivated garden shrubs (either by direct consumption or from discarded pruned branches). Pieris is a more deadly member of the rhododendron family, and unfortunately for livestock, is now widely grown as an ornamental garden plant. In a case encountered by the author, one Pieris bush 18in high was consumed by three goats, two of which died rapidly. Clinical signs include lethargy, salivation and repeated swallowing, vomiting of the rumen content, and if sufficient material has been consumed, recumbency and death will follow. A common sequel in those that survive is inhalation pneumonia caused by ingestion of the rumen content. Prompt veterinary attention is required if rhododendron or Pieris toxicity is suspected.

Ragwort: Goats seem to be fairly tolerant to this plant, which is rarely consumed in its fresh state. Poisoning can theoretically occur, however, if ragwort is incorporated in hay/silage, and care should be taken when selecting fields for forage conservation. Its

Diarrhoea	Sudden death	Nervous signs	Constipation	Stomatitis – mouth lesions
Hemlock	Yew	Ragwort	Oak	Giant hog weed
Oak	Laurel	Horse tails	Ragwort	
Water dropwort	Foxglove	Hemlock		
Box	Water dropwort	Water dropwort		
Rhododendron	Blue green algae	Black nightshade		
	Pieris	Male fern		
		Rhododendron		
		Laburnum		
		Laurel		

Haemorrhage	Frothy bloat	Anaemia	Vomiting	Discolored urine
Bracken	Legumes, e.g. clover	Rape	Rhododendron	Bracken
		Kale	Pieris	Oak
			Black nightshade	Rape
				Kale

Clinical signs of plant poisoning.

These yew trees have been topped – where have the clippings gone – could your goat find them if you live next door?

toxic principal damages the liver, and signs include depression, inappetance, emaciation, incoordination and jaundice.

Chemical Poisons

Although not common, there are occasional reports of chemical poisoning in goats. Clinical

These Pieris leaves were found in the rumen content of a goat at post mortem.

signs can often be rapid and severe, and an understanding of the potential sources should enable the reader to protect their goats.

Lead: This is one of the most common causes of chemical toxicity in farm animals. Most modern products contain very little lead but older materials such as paint or putty can contain dangerously high levels. Take care when using old doors or window frames as barriers, since any flaking paint or crumbling putty could be very high in lead, and may well be consumed by the inquisitive goat. Old, buried paint cans can occasionally come to the surface when ditches are being dug or trees are being pulled out and, yet again, goats may investigate anything 'new and unusual' in their environment. Broken car batteries are also high in lead.

Clinical signs include blindness, head pressing, disorientation and abdominal pain. Your vet may take blood samples from a live goat or kidney samples from a dead goat at + to check the lead status. Treatment in the early stages can be effective, but your vet needs to be contacted urgently if such treatment is to be successful.

Copper: Copper deficiency is referred to in Chapter 10, as a cause of swayback in newborn kids, and safe ways of supplementing were given. There is, however, a danger of over supplementation with copper to prevent swayback, resulting in copper poisoning. Goats are considered to be more resistant to copper toxicity than sheep, but care must still be taken with:

- mineral licks/blocks with added copper;
- feeds known to be high in copper, for example, pig rations;
- giving oral copper sulphate as a drench;
- footbaths containing copper sulphate, as these may be drunk, particularly if they are not carefully disposed of.

Excess copper taken in via the diet or other source will initially accumulate in the liver, but eventually a critical point is reached when maximum copper levels are exceeded, and the copper is released into the bloodstream. What is then referred to as a 'haemolytic crisis' will follow, in which massive rupture of the red blood cells occurs.

Clinical signs include anaemia, jaundice and abdominal pain, although affected goats may simply be found dead.

Metaldehyde: Main constituent of slug pellets. Poisoning by slug pellets, although a rare occurrence, has been encountered by the author on one occasion. A bag carrying slug pellets had split on the way to a field before being spread, and pellets had been scattered on the grass where a number of goats were grazing. Only one goat was affected, and was very disorientated for twenty-four hours, but it did make a full recovery.

If you suspect your goat has been poisoned:

- If you have identified the possible source of the toxin (either a poisonous plant or chemical) make sure you separate the goat from the source. If the goat is actually chewing the plant (for example, yew), you may be able to remove the plant material from the goat's mouth – but remember goat's teeth are very sharp!
- Stay with the goat, keep it moving if possible, and contact your vet as a matter of urgency with information on what type of poisoning you think may have occurred. There is evidence to suggest that dosing the goat with large quantities of strong tea may be useful, particularly if you suspect

a chemical poisoning; however, do not give tea if you suspect acorn poisoning, since both contain tannic acid!
- Your veterinary surgeon will almost certainly give only supportive treatment, since there are very few specific antidotes that can be used if poisoning is suspected. Your vet will attempt to control specific symptoms such as convulsions or spasms by sedation, and may administer fluids or vitamins to control shock. Antibiotics are often given if there is a danger of inhaling rumen content (a sequel to rhododendron poisoning).
- One further option is a rumenotomy, when an incision is made in the flank of the goat to gain access to the rumen; then the contents can be removed, particularly if plants have only recently been consumed.

There are some plants that can cause a noticeable taint or odour to goat's milk if eaten – although they may also be toxic if eaten in larger quantities. The following list gives a few examples:

Fool's parsley	Onion and garlic
Beet	Turnips
Shepherd's purse	Hemlock
Cress	Ivy
Henbane	Horsetails
Some clovers	Mint
Pea	Bracken
Oak	Buttercup
Radish	Yew
Laburnum	

CHAPTER SEVENTEEN

Zoonoses

Zoonotic diseases, or 'Zoonoses', are defined as those diseases that are capable of being transmitted between animals and man, although in reality it is mainly humans that catch such diseases from animals, rather than vice versa. It is important to emphasize from the outset that there is a group of diseases that can be picked up from goats by those handling them or in close proximity to them. Some are readily recognizable and can thus be avoided (for example, orf), but there are others in which the goat can appear to be totally fit and healthy, yet can act as a source of infection (for example, *E. coli* O157). Anyone handling goats should be aware of this risk, which can be conveniently broken down into three components:

1. Risk assessment.
2. Risk communication.
3. Risk management.

Risk assessment: All goats pose a potential risk to anyone handling them, having contact with the environment in which they live, or consuming products derived from the goat. However, this risk can vary from one that is real, to one that is minimal and inconsequential (depending wholly on local/individual circumstances). Those humans most at risk are the very young, the elderly, and those whose immune system is compromised due to ill health or specific medication – for instance, for cancer. Individuals within these categories should be kept away from goats, or protective barriers should be created between individuals and goats where possible.

All others having contact should consider themselves at risk, although observing normal hygienic procedures will help: this includes washing hands after handling goats, refraining from eating or smoking while handling goats, and wearing gloves for particularly unpleasant procedures. Equally, an assessment of the risk itself can be made; thus a fit healthy goat with normal pelleted faeces will pose less of a

The importance of thorough hand washing after handling animals should be taught at an early age (my daughters Maddy and Jessica).

Jessica feeding a goat on an open farm (left), but immediately adjacent to the picnic area (right).

risk than, say, a kid with diarrhoea (possibly Cryptosporidia), or a kid with orf lesions, or an adult with ringworm.

Risk communication: It is important that any risk identified (and your vet may well point out such a risk to you) is then relayed to anyone else having contact with the affected goats. This may be family members, a farm

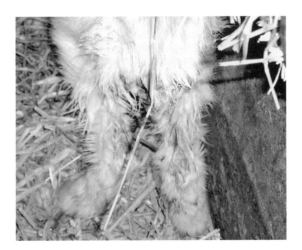

This goat kid has diarrhoea; it is therefore sensible to keep young children away, as there is a greater risk of zoonoses such as Cryptosporidia.

workforce, or perhaps, and most importantly, visitors to the farm such as school children, friends or neighbours.

Risk management: This revolves around your approach to any problem identified, but at the very minimum should ensure that there are adequate hand-washing facilities near goat accommodation, ensuring that all those having contact with goats know where they can be found.

The remainder of this chapter will consider some specific examples of known zoonotic diseases and their management.

ZOONOTIC ORGANISMS OF ENTERIC ORIGIN

Cryptosporidia

Already described in Chapter 6, this condition is caused by a protozoal parasite, and is normally associated with diarrhoea in young kids. Children appear to be particularly susceptible to cryptosporidiosis, and human incidents are occasionally linked to visiting groups of young children handling infected

'Fun hand-washing facilities', seen in a children's farm corner.

kids when bottle-feeding. It is important that kids with diarrhoea are kept away from such visitors, and that hand-washing under parental or adult supervision is carried out, with conveniently sited hand-washing facilities. Those humans who are affected will develop severe diarrhoea with abdominal cramps.

Salmonella and *E. coli* O157

These are two further enteric organisms that can also be passed on to humans (particularly children) in a similar manner to Cryptosporidia. Goats affected with Salmonella will often (but not always) have diarrhoea; *E. coli* O157, on the other hand, can be excreted by goats showing no signs of illness and passing normal faeces. Control again is based on assessing the risk, and ensuring that normal hygiene measures are followed. Human infection with *E. coli* O157 can be a serious problem since, in addition to diarrhoea, children can develop complications, with renal failure a potential sequel.

ZOONOTIC ORGANISMS LINKED TO SKIN DISEASE

Contagious Pustular Dermatitis: 'Orf'

(*See* Chapter 12.) This condition is seen most commonly around the mouth and lips of young

kids, and can be readily picked up by anyone handling them. Lesions in humans will tend to occur on the hands, but can be spread to lips, ears and eyes if these are inadvertently rubbed. As in goats, they begin as blisters.

Ringworm

Ringworm can be picked up by humans handling infected goats (*see* Chapter 12), or by handling fixtures and fittings contaminated with the ringworm spores that can be resistant in the environment. Lesions are initially itchy and red, but quickly develop a circular shape with skin crusting. They begin on the hands or arms (particularly in areas where clothes may rub), but can be easily spread to the face by rubbing.

Caseous Lymphadenitis

Although not strictly a zoonosis, there are reports of infection being picked up from discharging abscesses, particularly through broken skin such as cuts, grazes or scratches. Infection will spread along the lymphatics to the nearest local lymph node, which will become enlarged and painful.

ZOONOTIC ORGANISMS LINKED TO ABORTION

Refer to Chapter 5, in which the risk to humans from a range of zoonotic organisms is discussed. Remember this applies not only to those people working or living with the goats, but also to any visitors, particularly the general public. Ensure that signs are clearly displayed if you allow visitors to observe your goats at kidding.

ZOONOTIC ORGANISMS LINKED TO THE CONSUMPTION OF MILK AND MILK PRODUCTS

Small herds of goats supplying milk for the owner's household do not pose any risk to

A sign displayed prominently, warning pregnant women of the potential dangers of some infectious agents causing abortion.

the wider community; however, any milk supplied from unregistered and unregulated holdings is clearly a potential risk to public health, including the owner and his immediate family!

The standards required on a goat unit have to be based on two assumptions:

1. Any milk could reach consumers unpasteurized, either intentionally, or through accidental process failure, or post-pasteurization contamination.
2. Any dairy animal could, at any time, be excreting organisms potentially harmful to health.

The milking process itself must be hygienically carried out, the udder and teats should be free of faecal contamination, and the hands of the milker must be clean. Avoid including milk from goats with a vaginal discharge. Any abnormal milk should be discarded. Milk should then be rapidly cooled, and stored in hygienic containers. In the UK, further advice can be obtained from the Food Standards Agency/Dairy Hygiene Inspectorate.

Notifiable Diseases in the UK

Notifiable diseases affecting goats in the UK	Clinical signs	Last reported in the UK
Anthrax	Sudden death – mainly affects cattle, rare in goats.	2002 (cattle)
Blue tongue	Affects goats and sheep – insect-borne, results in high fever with high mortality.	Never
Brucellosis (*Brucella melitensis*)	Affects sheep and goats, will result in abortion storm if introduced into a group of goats.	1956
Contagious agalactia	Affects sheep and goats, causing mastitis, change in milk consistency and loss of milk supply, with other generalized signs of illness.	Never
Foot and mouth disease	Affects goats and all other cloven-hoofed animals. A highly infectious disease causing devastating outbreaks worldwide. Not usually severe in goats, but they develop typical blisters or vesicles in and around the mouth, the feet and udder and teats.	2001
Goat pox	A viral infection, initial signs are rapid onset of fever, salivation and nasal discharge and conjunctivitis. Skin lesions erupt in a few days. These develop into vesicles, followed by pustules and scabs.	Never
Pest des petits ruminants	Occurs in sheep and goats, and is referred to as Goat Plague in the latter. Mucosal erosions appear, stimulating profuse salivation and severe diarrhoea. Nasal and lachrymal discharges become mucopurulent and encrust, blocking the nostrils and cementing the eyelid together. Pneumonia adds to the distress of the animals. Both the breath and fluid faeces are disgustingly foetid.	Never
Rabies	Dogs and other mammals.	1970 (dog)
Rift Valley fever	Mainly affects sheep and cattle, rare in goats.	Never
Scrapie	Affects sheep and goats – *see* Chapter 10.	Present

Suspicion must be reported to the Government Veterinary Service (Defra). For further information in the UK visit Department for Environment, Food and Rural Affairs (Defra) at: www.defra.gov.uk.

APPENDIX II

Goat Weights

Weight

Adult Dairy	doe	55–105 kg
	buck	75–120 kg
Adult Angora	doe	33–55 kg
	buck	85–70 kg
Adult Pygmy	doe	22–27 kg
	buck	28–32 kg

Weight can be estimated by measurement of the heart girth (*see* Table 1). For best results, the goat should stand square with its head erect and the tape should fit snugly around the circumference of the goat, just behind the front legs. A tape is more accurate with dairy goats than with Pygmies. Age is related to weight by a normal growth curve (*see* Tables 2 and 3). Estimating the weight of immature Pygmy goats is not easy. Table 3 gives very approximate weights that can be expected at different ages. Females are considered mature at 24 months, males at 30 months.

Note: *See* Appendix III for metric and imperial measurements' conversion figures.

Girth (cm)	Dairy goats Weight (kg)	Pygmy does Weight (kg)	Pygmy males Weight (kg)
30	2.75		
35	4		
40	6		
45	9		
50	12	9	9
52.5	14	10.4	10.4
55	16	12.2	12.2
57.5	18	14	14.5
60	21	16	17
62.5	23	17.6	19
65	25	19.2	21.4
67.5	28	21	23.4
70	30	23.2	24.8
72.5	34	26.4	28
75	36	28.6	30
77.5	40	31	33.2
80	43	33.6	35.6
82.5	47	35.6	38
85	51	37.4	40.4
90	60		
95	69		
100	78		
105	88		

Table 1   Goat weight table

Age (months)	Weight (kg)
Birth	4
1	11.5
2	18
3	25
4	29.5
5	34
6	38.5
7	42
8	45.5
9	50
10	52
11	54.5
12	59
18	70
24	77
30	81.5
36	93

Table 2   Age-to-weight relationship (Dairy goat)

Age (months)	Weight (kg)	
	Male	Female
Birth	1.55	1.4
1	4.8	4.3
2	8.3	6.8
3	11.8	9.1
4	13.2	11.1
5	14.5	12.8
6	15.5	14.7
8	19	17.2
10	21	19.5
12	22.7	22.2
24	29	25
36	32	28

Table 3   Age-to-weight relationship (Pygmy goats)

Weights and Measures

Length

1 metre	=	39.4 inches	3.3 feet	1.1 yards
1 yard	=	0.9 metre		

Area

1 hectare	=	2.5 acres
1 acre	=	0.4 hectares

Volume

1 litre (1,000 millilitres)		=	1.8 pints	0.2 gallon
1 gallon (8 pints)		=	4.5 litres	
1 pint (20 fluid ounces)		=	568 mls	

Weight

1 kilo (1,000 grams)		=	35.2 oz	2.2 lbs
1 lb (16 oz)		=	454 grams	

Sample weights of a l litre jug of different feeds

1 litre of pelleted feed	Weighs approx 625 grams
1 litre of soya meal	Weighs approx 600 grams
1 litre of crushed oats	Weighs approx 400 grams
1 litre of coarse goat mix	Weighs approx 350 grams
1 litre of dried sugar beet shreds	Weighs approx 325 grams
1 litre of flaked maize	Weighs approx 250 grams
1 litre of bran	Weighs approx 150 grams

Corrected to 1 decimal place

Glossary

Abomasum The fourth or true stomach in ruminants; it secretes gastric juices, breaking down food material that has passed through the rumen.

Abortion Expulsion of the foetus/foetuses by a pregnant female before the normal end of pregnancy.

Abscess Formation of pus, usually within a capsule.

Acidosis A condition in which the pH of the rumen content falls, and the content becomes very acid. It is normally related to excessive intake of cereal, e.g. barley.

Acute (condition) A condition that develops rapidly with a sudden onset, and usually has a short course culminating in improvement or death.

Ad lib Any product feed or mineral mix, available for goats to eat to appetite – i.e. they decide how much they take in.

Afterbirth More correctly referred to as the placenta or 'cleansing'. It surrounds the developing foetuses in the uterus, containing the protective foetal fluids. When kidding begins, the membranes rupture and are normally expelled a few hours after kidding has finished.

AI (artificial insemination) The passage of semen (often diluted) collected from a buck, via a catheter (length of tube), into the uterus of a doe on heat.

Alopecia Loss of hair.

Anaemia Literally 'lack of blood'; may be related to blood loss, or an abnormality in red cell production.

Anaerobic bacteria Bacteria able to grow in the absence of air.

Anoxia Lack of oxygen – can develop if the airway is obstructed.

Ante mortem Before death/any procedure carried out in the live animal.

Anthelmintic A compound that kills or expels internal parasites such as worms.

Antibiotic A compound originally prepared from moulds (e.g. penicillin) but now manufactured artificially, that kills or suppresses the growth of bacteria. They can be administered by a number of routes, such as orally or by injection.

Antibody A protective protein produced by the goat in response to infection with any organism such as bacteria, a virus, or a parasite. Antibodies effectively result in the goat becoming protected against future diseases, and are also produced following vaccination.

Ascites Accumulation of fluid in the abdominal cavity.

Ataxia Incoordination of movement – a 'staggering gait'.

Atrophy Wasting away, or decreasing in size of cells, organs or entire areas of the body due to lack of use, disease or malnutrition.

Ballotment A procedure often used to detect pregnancy in the latter stages, when the kid(s) can be felt by gentle pressure applied to the abdomen.

Biosecurity A term often used to describe a range of procedures carried out to protect livestock on a farm against the entry of disease.

Bloat Accumulation of gas in the abdominal organs, usually in the rumen.

Body-condition score A numerical value given to assess the body condition of a goat. This book describes an approach based on two measurements over the lumbar area and sternum.

Bolus A large oval-shaped tablet containing, for example, antibiotic or mineral, that can be passed to the back of a goat's mouth and be swallowed.

Booster vaccination A dose of vaccine given to increase (or boost) levels of antibody to specific diseases.

Bottle jaw *See* oedema.

Breeding season The period of time the doe is on heat (in oestrous).

Browse The normal feeding behaviour of goats in the wild, whereby they feed on a wide variety of materials including grass and meadow plants, tree branches, hedgerow branches etc.

Buck/billy Uncastrated male goat of any age.

Butting Method of fighting among goats, especially bucks, by the striking of the head and horns.

CAE Caprine arthritis encephalitis, a viral disease of goats described in detail in this book.

Caesarean operation (or section) An operation to remove kids through the flank if they cannot be delivered naturally.

Calculi Stones that can develop at differing sites in the body such as the urinary system (*see* Urolithiasis).

Castration Procedure to remove the testicles.

Cauterize To burn or apply strong heat to an area, for example during disbudding.

Chewing the cud *See* cudding.

Chlamydia (Chlamydophila) The organism that causes enzootic abortion in goats and sheep. It is a zoonosis in that it poses a potential threat to humans in contact, particularly pregnant women.

Chronic condition A long-standing problem (as opposed to an acute condition).

CLA Caseous lymphadenitis, associated with the development of abscesses in many sites around the body, described in this book.

Clean grazing Land where goats (or other ruminants) have not grazed for some time, and is likely to be free of worm larvae.

Cleansing *See* afterbirth.

Closed herd One that does not buy in replacement stock, but rears its own replacements.

Clostridia Anaerobic organisms (grow best in the absence of air), widely distributed in farm environments, and considered to be normal inhabitants of the goat gut. *C. perfringens* is the most important member of the group, and can be associated with severe intestinal disease and death.

Cloudburst *See* pseudopregnancy.

Coccidiosis A protozoal parasite disease; infection causing damage to the intestinal lining such that normal digestion is inhibited. There is massive multiplication potential, such that the environment (the main source of new infection) becomes heavily contaminated.

Colic Abdominal pain.

Colostrum Milk produced after kidding by the doe prior to and during the first milking, which contains the immunoglobulins or antibodies.

Compound The portion of the ration fed that is usually high in energy and protein, and is balanced with vitamins and minerals. It may be in the form of a pellet or coarse mix (a blend of cereal grains and dried material such as crushed peas and beans).

Concentrate *See* compound.

Cornea The clear front part of the eye.

Coronary band The dividing line at the top of the foot, separating the hoof from the skin above.

Corpus luteum Referred to as the CL. This structure develops in the ovary after an egg is released and remains viable throughout the cycle. Secretes progesterone, and is responsible for the maintenance of the pregnancy.

CP (crude protein) *See* DCP (digestible crude protein).

Creep Can refer to the area in which young kids can seek refuge/safety from older animals;

and to creep feed, one fed in this area, exclusively for very young kids.

Cryptosporidiosis A protozoal (single-celled) organism that proliferates in the small intestine causing diarrhoea. It can cause illness in humans.

Cud/cudding The cud is food material regurgitated into the mouth for further chewing while the goat is resting; it is then re-swallowed and the procedure repeated. The whole procedure is referred to as 'cudding' or 'chewing the cud'.

Culling The process of removing goats that are below average in production, or are ill. Goats are occasionally culled on humane grounds if they are deemed to be suffering.

DCP (digestible crude protein) A term still found on the labels of feedstuffs purchased, but now replaced scientifically by RDP and UDP. It is a measure of the actual digestibility of protein when it reaches the rumen. If a large amount is degraded, it leaves only a small amount for true digestion.

Defra Department for Environment, Food and Rural Affairs – the UK agriculture authority.

Dehorning Procedure to remove the horns or terminate horn growth permanently.

Dental pad Goats have no upper incisor teeth, but instead have a firm pad referred to as the dental pad.

Dermatitis Inflammation of the skin.

Disbudding The practice of cauterizing with a hot iron the developing horn buds. The procedure must be carried out in the first seven days of life.

Doe A sexually mature female goat.

Drenching The oral administration of a liquid medicine (or drench) using a drenching gun.

Dry matter A feeding term used to compare the nutritive value of different feedstuffs, by expressing each one as a percentage with reference to its dry matter.

Drying off End of lactation when milking stops and the udder is allowed time to regenerate milk-producing tissue.

Dystocia Difficult birth.

EAE Enzootic abortion of ewes – often shortened to enzootic abortion or EAE – caused by Chlamydophila infection.

Ectoparasites Parasites that live on the skin surface, or that burrow into the skin, e.g. lice, mange mites.

Elastrator The instrument used to apply heavy rubber bands around the base of the scrotum to result in castration. Only a skilled person must carry out this procedure.

Emaciation A condition in which the goat progressively wastes away; it may be related to disease, or to failure to feed or find feed.

Encephalitis Inflammation of the brain; causes a variety of clinical signs depending on severity.

Endoparasites Parasites that live inside the body, e.g. worms and fluke.

Enteritis Inflammation of the gut, usually results in diarrhoea.

Enterotoxaemia A term usually applied to infection with *Clostridium perfringens*, in which toxin builds up in the gut, and spreads round the body causing an overwhelming toxaemia – often fatal. May be referred to as 'Type D Enterotoxaemia' specifically involving *C. perfringens* Type D.

Entropion A congenital condition in which the eyelid is rolled inwards, causing the lashes to rub on the eyeball causing severe irritation, pain and damage.

Enzootic abortion *See* EAE.

Epistaxis Nose bleed.

Euthanasia Putting a goat to sleep by a humane method.

External parasites Ectoparasites (*see* above).

False pregnancy *See* pseudopregnancy.

Fasciolosis Liver fluke infestation.

Fecundity Term referring to the number of kids actually produced by a doe.

Fleece Mohair or cashmere shorn from the goat.

Foetus The developing kid in the latter stages of pregnancy.

Follicle A cyst-like structure that develops in the ovary, resulting in the release of an egg or ovum.

Footbath A shallow container through which goats can be walked, and usually containing a chemical such as zinc sulphate, used to treat/control foot problems.

Forage Bulky feedstuffs such as hay, grass or silage.

Gangrene Death of body tissue in a live animal as a result of disease or an interrupted blood supply to the area.

Gestation The time between conception and kidding, usually between 142 and 152 days.

Gid A tapeworm cyst, usually in the brain.

Haematoma Blood blister – can occur in the ear flap, for example.

Haemonchus contortus The 'barber's pole worm' – lives much of its life cycle in the abomasum where it sucks blood, causing anaemia.

Heat *See* oestrus.

Helminths The main group of worms found in the gut.

Hypocalcaemia Low blood calcium levels – often encountered in does around kidding.

Hypoglycaemia Low blood glucose levels – can occur in young kids.

Hypomagnesaemia Low blood magnesium – rare in goats.

Hypothermia Chilling – a problem in young, newly born kids as the body temperature falls.

Hypoxia Deprivation of oxygen.

Immunity The development of resistance against a specific pathogenic organism (bacteria/virus/parasite), usually by the production of antibody.

Impaction Blockage – usually in the gut, caused by feed or faeces.

Interdigital space Area between the claws on the same foot.

Intermediate host An animal or other living thing through which a parasite may pass to complete its life cycle. The best example is the snail that is the intermediate host in the life cycle of liver fluke.

Internal parasites *See* endoparasites.

Intramammary tube Antibiotic solution inserted into the udder via the teat end.

Intramuscular Or I/M injection into a muscle.

Intravenous Or I/V injection directly into the bloodstream via a vein.

Iodine A chemical element vital for life, a deficiency of which causes goitre. Tincture of iodine (often referred to as veterinary iodine) is however manufactured specifically for application to the navel of a newborn kid after birth. It will rapidly dry the navel, causing it to shrivel, and thus preventing micro-organisms gaining entry to the body. Tincture of iodine should never be given by mouth.

Jaundice A condition, usually of the liver, causing a yellow coloration to the mucous membranes.

Johne's disease A wasting disease of goats caused by *Mycobacterium paratuberculosis*.

Kerato-conjunctivitis Inflammation of the eye, specifically the cornea and conjunctiva.

Ketones/ketosis Ketones are produced in the blood of a goat that is in a negative energy state (through starvation or pregnancy) when it begins to break down its body fat. Ketosis is the condition that develops, and needs urgent veterinary attention.

Kidding The act of giving birth to a kid (baby goat) by the doe.

Lactation The period between kidding and drying off when the doe produces milk.

Leukocytes Usually referred to as 'white blood cells', and part of the body's defence against disease or injury.

Liver fluke A parasite with a two-host life cycle involving the goat (and other farm animals) and a snail as the intermediate host. Fluke causes damage to the liver where it develops and grows.

Lochia The dark discharge a doe has from its vulva for a number of weeks after kidding.

Lungworm Roundworms (nematodes) found in the airways and lung tissue, causing damage and irritation.

Luteolysis The process whereby the corpus luteum regresses either as the next heat approaches, or at the end of pregnancy to begin the birth process.

Mange mites Small parasites that live either on, or burrowed beneath, the skin.

Manure The waste material (faeces) produced by the goat; when cleared out from housing it is generally mixed with straw, or other, bedding, and is then heaped up ready for spreading on the land.

Mastitis Inflammation of the mammary gland (udder).

Maternal immunity Antibodies produced by the dam giving protection against certain diseases that can be passed to the kid via colostrum.

Meconium Foetal faeces – yellow and sticky, passed out in the first few days of life.

ME (metabolizable energy) The amount of energy in a feed material that is available to the goat after the food has been digested. The unit of ME is the megajoule (MJ or 1 million joules).

Metritis Inflammation of the uterus/womb that usually follows kidding.

Milk fever The clinical picture produced by a low circulating level of calcium in the blood (hypo-calcaemia). The condition occurs most commonly around kidding, and may vary from unsteadiness on the feet to recumbency and coma.

Milk replacer Artificial milk substitute mixed with water and fed to kids to rear them artificially.

Mineral/mineral mix Term used to describe a range of inorganic substances that goats need for normal body function, e.g. sodium, chlorine, copper, magnesium. They are normally supplied ad lib.

Morbidity The number of susceptible goats that are affected during an outbreak of disease, usually expressed as a percentage.

Mortality The number of susceptible goats that die during an outbreak of disease, usually expressed as a percentage.

Mucous membranes Inner lining of cavities and hollow internal organs of the body, e.g. mouth, eyelids, nose, vagina and uterus. They secrete mucus.

Mucus Slimy liquid produced by cells' lining, e.g. vagina and other mucous membranes.

Mycotoxin Toxic compounds produced by moulds and fungi that can build up in spoiled feed.

Necropsy *See* post-mortem examination.

Nephrosis/nephritis Conditions affecting the kidneys.

Neurological Any condition affecting the nervous system.

Nodule A small discrete lump.

Notifiable disease A serious disease that is reportable by law (*see* Appendix I); there is a list of them, and they may vary around the world.

Oedema A collection of fluid developing in the tissue (often under the skin), e.g. 'bottle jaw', in which oedema fluid builds up under the skin as a result of, e.g. fluke.

Oesophagus The food tube or gullet connecting the mouth with the stomach.

Oestrogen Hormone produced by the doe that stimulates oestrus (heat).

Oestrus The period of sexual receptivity in the doe.

Oestrus cycle Length of time from one period of oestrus to the beginning of the next; averages eighteen to twenty-three days.

Omasum The third part of the goat stomach.

Oocyst A stage in the life of protozoal parasites such as coccidia and cryptosporidia that are shed in the faeces of an infected goat. These oocysts are then capable of causing disease in susceptible goats who pick them up from bedding or when grazing.

Oral Related to the mouth – e.g. an oral medicine is one given by mouth.

Ovary The primary organ of reproduction in the doe.

Palatibility Essentially how 'tasty' a feed is – a goat will readily eat a palatable food, but may refuse to eat one that is unpalatable, e.g. through spoilage.

Parasite An organism that lives a portion of its life cycle in or on the host.

Parturition The process of giving birth, kidding.

Pathogen Biological agent (e.g. bacteria, virus, nematode, protozoa) that produces disease or illness.

PGE (parasitic gastro-enteritis) Diarrhoea and gut damage caused by worms.

Pink eye A highly contagious disease of goats (and other ruminants) resulting in severe damage to the front of the eyeball.

Pituitary gland A gland situated in the skull on the underside of the brain, producing hormones influencing many organ systems in the body.

Placenta *See* afterbirth.

Post-mortem examination An examination of a dead goat to ascertain the cause of disease or death. To be effective, the examination must be undertaken as soon after death as possible.

Post partum After birth – i.e. any event that occurs after kidding has taken place, such as post-partum bleeding from the vulva.

Pre partum Literally 'before birth'.

Pregnancy toxaemia A metabolic disease of pregnant does generally thought to be related to a deficiency of energy in the later stages of pregnancy. *See* ketones/ketosis.

Premature When referred to kidding, it is the birth of a kid that is before its normal time.

Primary vaccination The first injection of a vaccine that sensitizes the body, ready for the second dose that results in the development of complete immunity.

Progeny Word referring to the offspring produced.

Progesterone A hormone produced by the corpus luteum that is dominant between oestrus cycles and during pregnancy.

Prolapse Pushing of abdominal organs through an orifice, e.g. vaginal and uterine prolapses.

Prolificacy *See* fecundity.

Protein Nitrogen-based essential nutrient, composed of chains of amino acids, that is present in all complex living things.

Pruritus Itching.

Pseudopregnancy Often referred to as cloudburst/false pregnancy. The doe ceases to come on heat, may develop swelling of the abdomen and udder development as if pregnant, but the uterus is empty. It may result in the sudden release of a large amount of clear fluid from the vagina.

Puberty Age at which a goat reaches sexual maturity.

Quarantine Period of time when a goat is kept in isolation away from other goats, to ensure that it does not introduce new diseases into the unit.

Ration A calculated mixture of feedstuffs fed to a goat on a daily basis, dependent on its demands such as lactation, pregnancy or growth.

RDP (rumen-degradable protein) The amount of protein degraded and absorbed from the diet in the rumen. Usually forage.

Rehydrate The replacement of fluids either orally as a drench, or by stomach tube or intravenous drip, to rectify a shortfall of fluid arising from disease or other illness.

Reticulum The second stomach of the goat located at the lower front end of the rumen. It is into this compartment that stones and other sharp objects will drop to prevent them moving on into the more delicate abomasum and lower bowel.

Rigor mortis The contraction of muscles causing rigidity that occurs after death.

Ringwomb Failure of the neck of the cervix to open correctly during the birth process.

Roughage Feeds high in fibre content and low in energy and protein digestibility.

Rumen The large fermentation compartment of the ruminant's stomach, whereby bacteria and protozoa break down fibrous plant material and other feedstuffs, and synthesize essential proteins and vitamins.

Ruminant Animals such as cattle and goats that ruminate and digest cellulose.

Rumination *See* cud/cudding.

Scald Infection of the foot causing soreness between the digits.

Scour/scouring A term often used to describe diarrhoea.

Silage Conserved grass or fodder, which is allowed to wilt and ferment in the absence of air. It can be either clamp silage (a large stack) or bagged/wrapped (smaller sealed and portable units).

Sire The father in a mating.

Sub-clinical disease Disease that develops but does not result in any obvious clinical signs (often referred to as asymptomatic or symptom free). Disease can, however, become clinical if it progresses, and illness may then be apparent. Sub-clinically infected goats may still shed infection.

Sub-cutaneous Or S/C: an injection under the skin.

Synchronization A procedure whereby does are brought into oestrus in a specific time frame, often by manipulation of day-length.

Trachea Windpipe.

Thymus A gland mainly in the chest cavity, which is large and active in young goats, and progressively shrinks in size. It is associated with the production of cells in the white cell series.

Total mixed ration (TMR) Complete ration consisting of concentrate, roughage and supplements necessary to meet the daily nutrient and energy nutritional requirements of the goat.

Trace elements Minerals that are considered vital for a healthy life, but are required in only small amounts.

UDP (undegradable protein) The amount of protein that is not degraded in the rumen, thus passing through for digestion in the abomasum. Concentrates are usually high in UDP.

Urolithiasis The development of calculi (stones) in the urinary system, a particular problem in males due to the anatomy of the urinary system.

Uterus Often referred to as the womb, the portion of the female reproductive tract where foetuses and kids develop prior to birth.

Vaccine/vaccination Suspension of altered or killed microbes or toxins (vaccine) administered to induce active immunity (vaccination).

Vaginal prolapse Protrusion of the vagina through the vulval lips in does in late pregnancy.

Vasectomy/vasectomized A surgical procedure carried out by your vet to prevent a buck being fertile, yet not interfering with his ability to mount and serve. Such a buck is used as a 'teaser' – i.e. he stimulates heat in does for service by other means including AI.

Viability The ability of an organism to survive.

Vitamins Small organic compounds necessary for proper metabolism that are found in small amounts in the feed. Deficiencies result in specific problems.

Weaning The removal of a kid from its dam.

Wether Castrated male goat of any age.

White muscle disease A disease caused by a deficiency of selenium and/or vitamin E, causing degeneration of the skeletal or cardiac muscle.

Withdrawal period The time scale during which a drug/medicine must not be administered if either meat or milk is to be marketed. A procedure designed to protect the human food chain from residues. Most products used will have a declared withholding/withdrawal period.

Wormer *See* anthelmintic.

Zoonosis/zoonotic disease Any disease that may be transmitted between animals and man. Examples are given in Chapter 17.

Index